GENERATIONS
OF THE MOON

GENERATIONS
OF THE MOON

JOHN QUINN

POOLBEG

Published in 1995 by
Poolbeg Press Ltd,
123 Baldoyle Industrial Estate,
Dublin 13, Ireland

© John Quinn 1995

The moral right of the author has been asserted

A catalogue record for this book is available from the British Library

ISBN 1 85371 309 0

Cover illustration by Roger O'Reilly
Cover design by Poolbeg Group Services Ltd
Set by Poolbeg Group Services Ltd in Garamond
Printed by Colour Books Ltd, Baldoyle Industrial Estate, Baldoyle, Dublin 13

For Katie and all . . .

and in memory of
Gordon Wilson
(1927– 1995)

THE PIHIS

Since first I read in *Zone*
what Apollinaire says of the pihis,
"They have one wing only and fly in a couple" –

I have heard their cries at midnight
And seen the shadows of those passionate
generations of the moon.

(Louis Simpson, Collected Poems,
Paragon House, New York, 1988.)

PART ONE

PROLOGUE

HANNAH

S HE TURNED HER HEAD SLOWLY, VERY SLOWLY, TO ONE SIDE. SHE COULD FEEL a ball-joint grating somewhere in the back of her neck. Pain burned across her eyes. She shut them tight and bit into her lower lip until the pain subsided a little. She released her eyelids slowly. She could smell a freshness from the sheets and feel a softness in the pillow. To a visitor the room would have looked soothing in its colours and airiness but Hannah Barden could find no such soothing as she battled to stem the tide of pain that surged upwards through her body and crashed on the rocks of her brain.

Her eyes focused on the bowl of flowers on the window-sill. A faint aroma drifted towards her from the flowers. She closed her eyes momentarily and savoured the aroma. Moira really did try hard to give her ease . . . Another wave dashed against her skull, causing her eyes to shoot open so wide that it would seem they must fall out of their sockets. Slowly the room came back into focus. Her eyes rested on a single eggcup with a rose design that stood just to the right of the flowers. In the eggcup rested a very white eggshell with an oriental design painted starkly in black on it. Moira had done that when she was ten, eleven maybe. A friend had taught her how to suck an egg dry by making a tiny perforation in the shell. She had then painted Japanese symbols on the shell and proudly presented it to her mother as a birthday present. Why Japanese she never knew

(Moira now laughingly described it as her early Japanese period) but Hannah treasured that present through the years. It had always occupied a prominent place in her bedroom. When she looked at it now, it assumed a new significance. She was that shell, drained of all the fluids and marrow of life. Within the shrunken shell of her own body everything grated and ground so excruciatingly that the slightest movement of any part of her body (like her head, just now) was tortured. And then in her mouth was that terrible dryness that no drink could dispel. A bitter rasping dryness. It reminded her of biting into an unripe sloe . . .

"What's them, Aunt Rose?"

"My but you're the curious wee lassie, Hannah Johnston!" Rose laughed. "What's this? What's them? Who's thon? You're non-stop with the questions. I have a question for you, young lady. Where do we live?"

"Sloe Hill," the child blurted with assurance.

"That's right – and these are sloes." Rose proffered a handful of the small blue-black berries to the child. "Go on. Take one. Just mind out for the wee stone inside."

Hannah chose the biggest berry and looked up to her aunt for reassurance before popping it in her mouth. Rose was nodding and smiling.

"Go on! Bite it!"

Hannah bit into the hard fruit. She winced as the acid taste reached her tongue. "Ach! Aunt Rose!" She did a little dance, shaking her head violently before spitting the sloe out into the long grass. "That's – awful!" she stammered as she tried to wash the tart coating from her teeth with her tongue.

Rose laughed heartily at the child's reaction and then quickly moved to ease her discomfort. She searched among the briars and plucked a few blackberries, "Here. These are sweet. These will make it better."

Hannah looked at her with distrust.

"Blackberries. Lovely and sweet. They'll take away the taste. Look!" Rose popped a blackberry in her mouth. "Grand and juicy!" she mumbled. Hannah took two blackberries and inspected them closely.

"They're nice and sweet. They really are!" Rose said. She ate another one herself. The child slid one blackberry very cautiously into her mouth, savoured it and then quickly added more. A smile returned to her stained lips. "Nice!" She nodded. "But them sloes – not nice." She winced as she remembered the sour taste.

"All right. Aunt Rose is sorry for playing a trick on Hannah." She ruffled the girl's blonde hair. "But you can help me pick sloes." She reached for the basket she had left in the grass when searching for blackberries.

The sloe bushes towered above Hannah who could only reach the occasional berry.

"Why are we picking sloes, Aunt Rose?"

"The questions are off again! Because we can make a very nice drink from them. You'll see later on." Hannah soon tired of the sloe-picking and sat in the tall grass watching a butterfly balance on a waving stem. She stared intently at the butterfly.

"Aunt Rose?"

"Mmm."

"What's a bloody Taig?"

Rose, leaning across the top of a sloebush, maintained her balance with some difficulty and gradually eased herself back to an upright position before turning to the girl.

"Where did you hear that, child?" She whispered the words in measured breaths.

"Winston," the girl replied nonchalantly, keeping her eye on the butterfly. "He was cleaning the horse's stable. He didn't see me. He just kept talking about bloody Taigs. Ah, come back, butterfly!" Hannah cried as the insect left its perch.

"Winston Pollock should learn to keep his mouth shut," Rose muttered to herself.

"So what's a bloody Taig, Aunt Rose?"

"It's nothing, child. It's just – other people. Come on. We've enough sloes for now . . ."

"Is that you, Aunt Rose?"

"No, Mum. It's only me – Moira," her daughter whispered, smoothing the crumpled sheet. "Can I get you anything?" She brushed her mother's hair back lightly with her hand.

3

"No, dear. I'm all right."

Moira watched her mother's hand scrunch the smoothed sheet. "Maybe a little drink?"

"Maybe."

She cupped her mother's head in one hand and held the cup in the other. Her mother's lips searched out the straw and she sucked, very slowly.

Moira drew a long breath and tried to sound casual.

"Brian rang today." Her mother's reply was a gurgling sound through the straw. "He says he may come over next week – to see you."

"Aye."

"He could get a flight to Heathrow and a train from Victoria."

"If he wants to waste all that money – "

There was a long silence, punctuated by the sound of Hannah's light sucking movements. When she had finished, Moira wiped the dribble from her chin and eased her mother's head back on the pillow. "Now," she said, again lightly brushing the wisps of grey hair, "don't you feel a bit better?"

"Aye." She waited and winced as another wave surged upwards though her body and crashed . . .

"A wee drop of sloe-gin always helps . . ."

Aunt Rose emptied the sloes into a huge colander and positioned the vessel under the pump in the yard. She pumped furiously until the water sluiced over the glistening fruit. "Have to wash any wee bugs and things out of them," she panted, anticipating Hannah's question. She picked up the colander and moved towards the kitchen. "Now, young lady, I need your help."

She sat the dripping colander on a tea towel on the table, took a large glass jar from the kitchen press and placed it beside the fruit. She took a pin-cushion from the mantelpiece and carefully withdrew two pins. She handed one of them to Hannah. "Now be very careful and do as I do." She held up a sloe, pierced it gently with a pin and dropped it into the glass jar. Hannah found it more difficult to pierce the blue-black skin of the sloe but she persisted and held up the fruit, impaled on the pin, in triumph. "Good girl," Rose beamed, "we'll make a grand wee cook of you yet!"

4

"Why are we sticking pins in the sloes, Aunt Rose?"

"To let the juice flow out, so we can make some nice sloe-gin for Christmas."

"I like Christmas."

"Mmm."

The two fell silently to work, the steady ticking of the clock on the mantelpiece interrupted only by the soft plop of a sloe dropped into the jar.

"Aunt Rose?"

"Mmm."

"Are you going to marry Winston Pollock?"

Rose dropped her hands flat on the table.

"Dear heaven, child. Where do you get these questions?"

"Well, are you?"

"I don't know, child," Rose replied tetchily, "I – I have no plans."

"Well if you do, will you still be my mammy?"

Rose shoved back her chair noisily along the flagged floor, rounded the table and swept the child out of her chair and into her arms. She held Hannah very close and ran her fingers through her blonde curls.

"Of course, child, of course." She stifled a sob as she made her way to the mantelpiece. She held the girl in her left arm while reaching to take down a photograph in a silver frame. In the photograph a tall, lean dark-haired man smiled nervously. He was seated beside a standing woman. The woman was slim and frail, her piercing serious eyes gave her an almost frightened look.

"Why does Mammy look sad?"

"I don't know, child. Maybe – I don't know. What I do know is that when she died I promised your daddy I would look after you – for as long as he wanted. And I'll keep that promise!"

She looked from the child's eyes to the mother's eyes in the photograph. That piercing look. She pursed her lips, closed her eyes momentarily, took a deep breath and looked straight into those piercing blue eyes.

"Look, child. It's time you were told. You're a bright girl and you'll be off to school next summer where no doubt there'll be wagging tongues."

"What's wag . . . ?"

"When your mammy had you, Hannah, she had another baby as well. A baby boy. His name was – is – Brian. He's your twin brother –"

"Twin brother," the child repeated, still staring at the photograph, "twin brother –"

"That's right. You have a little brother."

"Where is he? Is he dead too?"

"No. He's not dead. He's away. He – he lives with – other people."

The kitchen suddenly darkened as a figure appeared at the door. His burly frame almost filled the doorway, the light above his shoulders threw his thickset features into menacing relief. It was Winston Pollock.

"Why don't you talk straight to the wee one?" he rasped, "why don't you tell her he was taken by the bloody Taigs?"

BRIAN

Brian Johnston cleared his throat, leaned his splayed fingers on the altar cloth and looked slowly around the little oratory. He counted, four, seven, nine. There were only nine people present. A young fresh-faced nun knelt beside the coffin. Sister Fiona, who had kept her promise and had summoned him when Aggie died. She smiled at him and sat back in the seat in anticipation. Further back three elderly nuns sat side by side, heads bowed, lips moving silently. Behind them two aged women each wearing an ill-fitting tweed overcoat. One raised and lowered her head continually, eyes closed, her fingers joining and unjoining in time with the rocking movement of her head. The other sat erect, smiling as she ran the fingers of her right hand up and down the buttons of her brown coat. To their right a middle-aged man, his stout frame straining the buttons on his dark blue porter's coat, leaned forward, resting his chin on folded hands. Two dark-suited undertakers stood, arms folded, on either side of the door. Nine people, Aggie. Why bother with a homily? Because it was necessary. Necessary to make amends. He gripped the altar cloth and began.

"We are gathered today to commend the soul of our sister Agnes to our Father in heaven and to celebrate her long life on this earth, over eighty years. It might not seem right to use the word 'celebrate' with regard to a life, the last fifty years of which were lived within this – institution . . ."

The woman in the brown coat now ran her left hand up and down the buttons.

"But in those fifty years Agnes – Aggie – was closer to God than any of us. Closer in her childlike innocence. Closer in the beauty and purity of her soul. Yes I know she was called – we called her – Mad Aggie. And Fatimaggie. We taunted her about having the third secret of Fatima. We ridiculed you, Aggie – all of us in whatever degree. Some of us should have known better. Should have known that you were – to use the lovely phrase from our own language – a *'duine le Dia,'* a person close to God . . ."

One of the elderly nuns blessed herself. Her two colleagues automatically followed her example . . .

"I say, Uncle James – "

"Aye."

"Why do you come up here to the top of the brae to say your prayers?"

"How did you know I was saying prayers?"

"'Cos Aunt Bridie told me."

"Did she now?"

"And she told me I mustn't annoy you."

"She did!" James burst into laughter and tousled the boy's hair.

"So why do you come up here – "

"I suppose it's so people won't annoy me!" James gently poked a finger into the boy's chest, tickling him until the boy doubled up in chuckles and backed away.

"And up here on top of the brae, amn't I nearer to God?" James added, gesturing towards the startling blue June sky.

"Is God really up there, Uncle James?"

"'Course. He's probably peeping out at you from behind yon wee cloud!"

The boy's gaze followed his uncle's outstretched hand. A wisp of cloud drifted away to his right.

"Yon's too wee a cloud," the boy announced with utter seriousness, "God's a big, big man."

"Do ye reckon?"

"Aye. Auntie Bridie says he's bigger nor any of us! I say Uncle James – "

"Aye?"

"Who's Jemmy Patchy?"

"Jemmy Patchy? My, but you're the curious wee man, Brian. Where did you hear about Jemmy Patchy?"

"From Podgy. Me and Aunt Bridie were in his shop yesterday and Podgy was very cross with the man that was working out in the shed. Podgy said he was as cracked as Jemmy Patchy."

"Trust Podgy to have a saying like that! He gives poor Hanratty a terrible time."

James closed his prayer-book.

"Jemmy Patchy used to walk the roads around here before you were born, Brian. He was a soldier one time, in the Great War, except no one ever told him the war was over! He used to march up and down the roads singing songs and pretending to shoot people."

"Did he have a gun?"

"Not at all. Only a big stick! And he wore a big long coat that was all tattered and patched. That's why he was called Jemmy Patchy. There wasn't a bit of harm in him though. Your mother was very fond of him. Many's the time she brought him down to the house for a bit of dinner. Which reminds me – is there any sign of dinner yet, I wonder."

"Aunt Bridie said it would be ready in about ten minutes," the boy replied. He began to march up and down, chanting the words "Jemmy Patchy" solemnly as he went.

James opened his prayer-book again.

"So while we ask our Father to forgive Aggie for whatever unlikely offences she committed in her long life, equally we ask you, Aggie, to forgive us. Forgive us for the hurt, for the long years of loneliness, for the sheer ignorance and pride which cast you from this world into the darkness and isolation of an institution. Our Lord spent forty days in the wilderness. Aggie spent fifty years there. Fifty years. More than a lifetime for many. And none of us asked why. Until it was too late. For this and more, Aggie, we ask your forgiveness –"

"Amen!" the woman with the rocking head movement cried aloud. Brian Johnston nodded towards Sister Fiona.

"We will now have the offertory gifts . . ."

"I say, Uncle James – "

"Yes, Brian."

"Why did no one ever tell Jemmy Patchy the war was over?"

James laughed and ruffled the boy's hair.

"Ach no, I didn't mean it like that! It's just that – Jemmy had a bad time in the war and he was a bit confused when he came home."

He plucked a stem of grass and drew it through his teeth.

"War can do terrible things to a man," James said quietly to himself.

"You can get killed in a war, can't you, Uncle James?"

"You can indeed, Brian – and many did."

"Were you in the war, Uncle James?"

"No I was too young for that war, thank God."

A figure waved to them from the house below.

"Uncle James, am I annoying you?"

James laughed loudly as he struggled to his feet.

"Indeed you're not, Brian. And look! Aunt Bridie has the dinner ready. Come on, I'll race you down the brae!"

He propelled the boy forward to give him a headstart and he hobbled in pursuit.

"That's not fair! I wasn't ready," James called in mock complaint.

The boy wasn't listening. He careered down the hill, arms outstretched, exulting in triumph.

"Dust you are and unto dust you shall return . . ." He sprinkled a handful of clay on Aggie's coffin, completed the burial prayers and withdrew to allow the grave-diggers begin their work of filling in the grave. The wind sighed through the yew trees of Cullyboe cemetery. The funeral had not come through the village, so little notice was taken of the hearse and the one car that followed it. Despite this a little knot of people had assembled at the cemetery gate. Most of them were elderly women who now joined Sister Fiona in a decade of the rosary for Aggie. They then melted away as quietly as they had come, one or two picking their way across the graves to visit a family plot. Brian Johnston followed their example and threaded his way through the uneven, often overgrown graves. At the end wall in the shelter of a very tall yew tree he found the grave.

Pray for the soul of
Manus McKevitt
Drominarney, died 10 April 1918
and his wife Kathleen, died 2 June 1920

The black lettering had disappeared totally from the marble headstone. Yellow lichens had crept into the grooves. He stooped to pull a weed that had grown up through the faded white chippings. The stem came away in his hand leaving the root intact. He tisked and sighed as he knelt on the grave-kerb. At the bottom of the headstone was a further inscription, more recently carved.

And their daughter Sarah,
died 21 December 1926

"Rest in peace now, Mother," he whispered, "rest in peace."

CHAPTER ONE

SARAH McKEVITT FIDGETED NERVOUSLY AS SHE WAITED IN THE HALL OF Cullyboe presbytery. She sat on a cushionless oak chair staring at the dark brown wallpaper on the other side of the hall. There were small patches of damp just above the black-stained skirting board. They probably accounted for the musty smell that pervaded the hall. High up on the opposite wall there was a glass case. She knew it was there but she did not want to look at it. It had been there twenty years ago when she came to this house while her mother was housekeeper to the priest for a short time. She knew what the case contained. A huge pike caught by Father McKenna in Lough Vinney. It sat there, ugly mouth agape, snarling at her. It had frightened her as a child. She still felt uneasy in its presence.

The parlour door opened and Miss Carroll, the housekeeper, appeared. "Canon McKenna will see you now," she said coldly. Sarah stood up and brushed past the figure in black, closing the parlour door quietly behind her. "She's probably bent to the keyhole already," Sarah thought.

Canon McKenna sat in a huge throne-like chair, his biretta resting on the desk in front of him. He rapped the bowl of his pipe on an ashtray and then blew through the stem several times. In between blows he motioned to the girl to approach him. "Come in, child. Take a seat." She sidled nervously into a high-backed plush red chair. "And how – " he sucked staccato at the pipe-stem, " – are all the McKevitts?"

"Very well, thank you, Canon." She managed a weak smile.

"A grand family. A grand family. Church and country are built on families like yours, Sarah. Never forget that, child. If the foundations are right, the edifice will stand. Isn't that right?"

"Yes, Canon."

"And the foundations in your case are a father and mother, God rest them, who responded so beautifully to their vocation. Hard-working, God-fearing, cherishing their faith and passing it on to their children by solid example – "

Sarah dug the nails of her right hand into her left palm.

"I want to marry Gordon Johnston, Canon."

"Solid example – Mass, the sacraments, high moral standards and the acceptance of God's will."

Her palms ached to the point where her eyes watered. She spoke softly and deliberately.

"I'm going to marry Gordon."

His eyes flashed at her as he brought the pipe down with a thud on the green baize, causing his biretta to bounce up from the desk.

"Ach, you're making a mistake, girl. You haven't thought this out at all."

"I've thought about it a lot, Canon." Her mouth was suddenly very dry and she had to force the words out.

"Look!" He shot an arm in the direction of a picture on the wall, a reproduction of the da Vinci cartoon of the Holy Family. "They are your models!" He swept his arm in an arc until it pointed directly at Sarah. She noticed the tremor in a nicotine-stained index finger. His voice, rising in pitch all the time, was now almost menacing.

"Joseph and Mary. Did Joseph go beyond his own when seeking a wife? No. Stay with your own, child. We've had too many divisions in this blighted country. The separation of six counties from the Free State. A civil war. And now," – the finger trembled even more – "you want to add to those divisions by marrying outside your faith –"

"He's a good man, Canon." She had disengaged her hands but the tears continued to well in her eyes. The Canon withdrew his arm at last and slowly brushed from the baize the ash which had spilled from his pipe. He lowered his voice.

"I have no doubt he's a good man, child. And that he will make a good husband and father – but to one of his own! For God's sake, girl, look around you. Aren't there plenty of young men in Cullyboe and beyond – good solid young Catholic men who would – "

13

She gripped either side of the chair. "My mind's made up, Canon." There was a long silence.

"We shall see. If only your good parents were alive this day . . . However, I'm warning you, child. It's a dangerous road you've chosen. And if you follow it, remember Mother Church will demand that your offspring be reared according to her laws."

"We've agreed that, Canon." Her voice was little more than a whisper.

"You're a very stubborn girl, Sarah McKevitt. I've done my duty. I've no more to say – for now. Good day!" He snatched at his biretta, stood up and swept out of the room.

Sarah wheeled her bicycle down the presbytery path towards the road. On an impulse she turned sharply down a minor path and through the small white gate that led to the adjoining cemetery. She negotiated her way through the narrow pathways of the newly laid-out graveyard, down to the back wall where her parents' grave lay in the shade of a young yew tree. She stood there silent, head bowed, hands resting on the handlebars of her bicycle. After a few moments she whispered aloud, "Please help me!", walked briskly out of the cemetery, adjusted her head scarf, mounted her bike and headed towards Drominarney.

The noon Angelus bell was ringing as she approached Drominarney Cross. Sarah smiled to herself. She had made good time. Instead of turning left for home, she went on straight. Two miles further on she came to another crossroads. Skirting the dancing flagstones on her left, she began the slow and winding climb up Caraher's Hill. Halfway up she dismounted and rested her bicycle against a rusting gateway. She stood on the lowest bar of the gate and scanned the field of recently mown hay. There was no sign of him. She was about to step down again when a low call came from beneath a hazel tree on her right. It was Gordon. She clambered over the gate and raced across the meadow, skipping over the newly-turned rows of hay.

"Well, isn't it well for some, having a picnic in the fields at this time of day?" She laughed, holding her right side which pained her after the exertions of cycling and running. She collapsed into his arms.

"Been here since eight o'clock this morning," he said.

14

"I know! I know! I was only joking!" She kissed him lightly on the cheek as he drank tea from a bottle.

"Well, did you see the Canon?"

"Mmm!" She lay back on the grassy bank.

"And?"

"Just warnings and threats on the danger of marrying a black Protestant!"

"And they accuse us of being a 'No Surrender' crowd – "

"I told him I was going to marry you – "

"As long as the children are on his side."

"Yes but we agreed that anyway."

"I know. I know. The Canon must have his pound of flesh – "

"Gordon!"

"I'm sorry. I suppose I'm just hot and tired. And I'm fed up of divisions in this country. I know by the look on old Caraher's face today that he'll not rent me this field any more." He stood up suddenly and walked to the nearest row of hay, idly turning it with the toe of his boot. Sarah watched with growing anxiety. He looked up the sloping field, his eyes growing narrow with frustration.

"What's the difference?" He turned to Sarah. "I mean what's the difference? They're the same wee fields, the same wee hills they were a few years ago, a hundred years ago. Until someone draws a line on the map and says that's one country and that's another! And suddenly people are different. They're cautious, suspicious. People you knew all your life." He shook his head in disgust. "It's just not right, not fair."

Sarah put her arms around him. "It will be all right, love. You'll see. Things will settle down."

"No. Things will never be the same. Bad enough me being of a different faith than you. Now I'm of a different country too."

"What are you saying, Gordon? Don't you want to marry me?"

"'Course I do. I just want you to know that in some ways it won't be easy for us, for you."

"God, don't I know that." She saw Canon McKenna's quivering nicotine-stained finger pointing at her. "I'd better go. Bridie will be wondering where I've got to. You won't forget the dance?"

"Dance?"

"Oh, Gordon! There's a dance in the schoolhouse tonight. It's to raise funds for the school. You will come?"

15

He was silent.

"It will be a chance – "

"To meet the opposition?" He smiled. His face changed totally. "I'll come."

She climbed over the gate. "You should smile more, Gordon Johnston. It suits you!" He reached across the gate, cupped her face in his hands and kissed her. She wriggled free. "Certainly does suit you," she laughed as she mounted her bicycle. "Nine o'clock – no later," she called back.

As she approached the Flaxpool Lane where she would turn down to her house, she noticed an unkempt figure shambling up the middle of the road ahead of her. She rang the bicycle bell continuously. "Ho there, Jemmy Patchy," she called, giving him plenty of warning. The figure turned suddenly and swung a huge staff from his right shoulder around into a "firing" position, crouching as he aimed the staff at Sarah.

Nobody knew who Jemmy Patchy was or where he had come from, but for years he had been a familiar sight along the roads around Cullyboe. Dressed in a tattered and patched greatcoat, under which he wore a ragged kilt and trousers, he marched the roads in boots that were far too big for him, carrying a long stout staff which became his gun when challenged. His piercing blue eyes shone out from the tangled matted mess that was his hair and beard. It was deduced from his behaviour and his dress that he was yet another piece of wreckage from the Great War, shattered in body and mind. He had been in the Dardanelles, if one could believe his occasional ranting. To a stranger he presented a frightening sight but in fact he was totally harmless. When he passed by the school yard at playtime, the children teased and taunted him but he would only laugh and sing a snatch of a song:

"If you want the Sergeant-major
I know where he is, I know where he is –"
When he was gone, the children bounced a ball to the tune:
"Oh the moon shines bright on Jemmy Patchy
His boots are cracking for the want of blacking
And his little baggy trousers they need mending
Before we send him to the Dardanelles . . . "

He fixed his stare on Sarah. "Bulgars?" he spat out the word. "Did you see any Bulgars?"

"Not a one, Jemmy. Not a one!"

He whipped his head around to the left. "No sightings, sir. Not a one."

Sarah smiled. "Will you take a bite, Jemmy?"

He looked to his left again. "Permission to fall out, sir?" He saluted, put the staff on his shoulder and stood aside to let Sarah lead the way. She wheeled the bicycle the last half mile to her house. Jemmy followed silently behind.

"I'm home!" Sarah called as she wheeled the bicycle into the barn.

"About time too," her sister shouted back, "fine time of the day to go *céilíng* – "

"And I have a guest for dinner!"

There was a clanging of saucepans. Bridie rushed to the door, wiping her hands on her wrapover apron. "Not Gor – , oh, 'tis yourself, Jemmy." The relief in her voice was audible. "Take your ease there, Jemmy, and I'll bring you out a sup to eat."

Jemmy saluted and sat on a mound of hay, clutching his staff across his knees. Bridie had no objection to bringing Jemmy into the house but she knew he would not stand being confined within four walls even for the length of time it would take him to eat a bit of dinner.

Sarah laid the table for dinner.

"Well?" Her sister over-busied herself with impatience.

"I got the sugar and the lard – "

"You know – "

"You know well what he said. I got a lecture on staying with my own kind, on following the example of Joseph and Mary, of our own parents."

"There's nothing wrong with that."

"There's everything wrong, Bridie McKevitt." She slammed a knife and fork down on the table. "You know I want to marry Gordon. God! You're as bad as the Canon!"

Bridie whipped around to face her sister but froze as she heard voices outside. Hughie and Pete were passing greetings to Jemmy Patchy. Bridie bent to ladle a bowl of stew from the pot that hung over the fire. She handed the bowl to Sarah. "Bring that out to your friend." The words were carefully measured. "Tell him to blow on it. It's hot!"

The men sat down to dinner, each balancing his cap on his right knee. The sisters joined them. There was little conversation.

"Much of a stir in the village?" Hughie enquired at last.

"Divil a bit. Podgy Kernan's aunt died, over in Newry Hospital."

"She'd be a brave age."

A long silence followed, punctuated only by the scraping of knives and forks on the plates.

"How's the hay going?" Sarah asked tentatively.

"Good crop," Pete mumbled through a mouthful of potato, "if the weather holds we'll have it all cut this evening."

"Gordon's already turning his!" Sarah announced with pride and then realised her mistake.

Bridie stood up to replace the pot with a kettle on the crook over the fire. Another silence. Hughie pushed away his empty plate.

"Pat Trainor's giving us a hand," he said casually, looking across at Sarah. "He said to be sure you're at the dance in the school tonight."

Outside the kitchen window Jemmy Patchy swaggered down Flaxpool Lane, his staff bobbing up and down above the hedge. His tuneless voice faded in the distance:

"Keep the home fires burning – HUP!

While our hearts are yearning – HUP!

Though our lads are far away – HUP!

They dream of home – HUP! HUP! HUP-HUP-HUP!"

CHAPTER TWO

THE DANCE IN DROMINARNEY SCHOOLHOUSE HAD BEEN ADVERTISED TO START at nine-thirty but it was a fine June evening and, as Sarah McKevitt walked to the school, the rattle of mowing machines echoing each other across the valley told her that the men, her own brothers included, would put hay cutting before any dancing. Many of them would also spend an hour or two in Podgy Kernan's pub before they would feel brave enough to ask a girl to dance.

The first person she met in the schoolroom was Canon McKenna. She averted her eyes and looked hopefully for somewhere to lay out her basket of scones and buns. "Ah Sarah!" he beamed at her, his red cheeks aglow in the light of his pipe. "Good girl! The McKevitts were always to the fore in helping their parish. I see you've been busy at the baking. Good girl! Good girl!" He took a bun from the basket. "Mmm! Delicious! Some young man is in for a treat tonight! A beautiful girl who can bake like that would make a grand home for a lucky man!" She half expected him to list the young men she might entertain. She smiled weakly but said nothing.

"And where is Bridie?"

"She had to do the milking, Canon."

"Martha, Martha, thou art troubled and anxious about many things," he mused. His voice trailed off as he moved outside to greet more arrivals. The desks had been moved out of one room to provide dancing space. They formed counters in the other room on which Sarah and others arranged their produce. Two large kettles sat on a turf fire in this room also. Sarah remembered the winter days as a child when she craved to be near that fire but shivered instead in a draughty corner. The crowd straggled in in twos and threes and it was well past ten o'clock before Matt McLarnon and his sister Josie struck up the music with melodeon and fiddle. The men fidgeted and laughed nervously on one side of the room, the women stood shyly alone or whispered in small groups on the other. Matt and Josie played away stoically in front of the fireplace. It wasn't until Jimmy McCabe was half pushed out to the middle of the floor that the ice was broken.

"Ah come on lads," he urged his comrades by the wall, "in for a penny, in for a pound!" He marched straight across the floor and took the wrist of the girl nearest him. The dance had begun.

Sarah busied herself in the tea room, helping with cups and arranging a display of cakes. She cast an odd anxious glance towards the door. It was when she had turned away from the door to talk to a neighbour that a hand was suddenly laid on her shoulder and a voice said, "May I have this dance?"

Her body went taut. It was Pat Trainor. "I – I'm sorry. I'm – busy here," she stammered.

"Nonsense, girl, off you go. There's plenty of us here." Bridie appeared from nowhere. The dance, an old-time waltz, was little more than an awkward shuffle on the now crowded floor. Pat Trainor smiled smugly, beads of sweat glistening on his balding forehead.

"You're making yourself horrid scarce these days."

"How's that?"

"You know well. You go out of your way to avoid me."

"That's your imagination, Pat."

"So you don't want to avoid me?"

"I – I didn't say that."

"At this dance last year, you danced nearly every dance with me."

"Last year was last year – "

"Before you met the Prod – "

She disengaged her arms from him but he locked his tightly around her waist. "I don't see him here tonight."

"He will be."

"I believe you were in with the Canon today."

"There's not much you miss."

"I hope he gave you good advice – "

"Look, Pat, the past is the past. I'm going to marry Gordon."

He relaxed his grip sufficiently for her to wriggle free and slip through the press of bodies. Pat Trainor was saved embarrassment by Matt McLarnon's announcement of a tea break.

Sarah fought her way to the tea room and clambered behind a desk to help serve refreshments to the dancers. She caught a glimpse of Bridie smiling across at her. Tears welled in her eyes. She grabbed a teapot from another woman. "I'll do that!" she whispered hoarsely. She kept her head down as she filled cups furiously, often sloshing the tea wildly onto the floor. Such was the clamour nobody seemed to notice her discomfort. The babble of voices and laughter reached a crescendo until all but the stragglers had been served. Gradually the crowd broke into knots and the hubbub descended into normal conversation on topics which ranged from politics to football.

"They say South Armagh will come into the Free State."

"Well if they do, we might get a football team out of it. Did you see them last week?"

"Bunch of oul' grannies!"

"Worst ever!"

Sarah emptied a kettle into a large aluminium basin and began washing cups. Matt McLarnon startled all present with a sharp burst on the melodeon and then called for a ladies' choice. Jimmy McCabe sleeked back his quiff and swaggered out on the dance floor. "Now girls, don't all rush! I'll not be responsible for deaths in a stampede!"

Sarah smiled and continued the washing-up. The floor gradually filled up with couples and it was at a very late stage that Bridie came to the rescue of a disconsolate Jimmy McCabe. Sarah gave a sigh of relief when someone swept Pat Trainor onto the floor but she still felt his eyes on her. She glanced again towards the door. Her heart jumped as she recognised a familiar shock of black hair. Gordon Johnston stooped to enter the schoolhouse. She dropped a cup back in the basin, carelessly wiped her hands on the sleeves of her cardigan and wove her way through the crowd to him. She grabbed his wrist and pulled him onto the floor. "It's a ladies' choice and you're my choice for the rest of the night!" She smiled up at his bemused face. "I thought you'd never come."

The McLarnons stopped playing at precisely one minute to midnight. Matt thanked everyone for coming and wished them safe home. The crowd poured out into the welcome cool of a June night. There was excited chatter, whoops of laughter and the odd ribald comment as bicycles were retrieved from hedges. Inevitably the voice of Jimmy McCabe rose above all others, " – the only conveyance here with a padded crossbar! No, Canon. I'm sorry. It's reserved!" There was immediate uproarious laughter.

"Mine's not padded," Gordon said, "but you're welcome nonetheless!"

"We'll walk. It's a lovely night." She linked his arm. As they sauntered down the moonlit road, the night air resounded with the calls and laughter of the departing revellers.

"It's nice to hear people so happy," Gordon remarked. "Maybe there's hope for us all yet."

"Maybe."

The furious ringing of a bicycle bell behind them caused them both to leap up on the grass verge. Jimmy McCabe sped past with, to Sarah's amazement, Bridie on the padded crossbar.

"Well done, Jimmy," Sarah called out in surprise. He waved one hand back at them. "The McKevitts always liked to travel in style," he shouted.

"That looks like the start of another romance," Gordon laughed.

"She could do worse than Jimmy." Sarah burrowed her head into his chest.

"Double wedding?" he jibed.

"Let's just worry about our own!"

He put his arm around her shoulder. They walked on in silence, breathing in the delicious scent of new-mown hay. The whoops and calls of the dancers gradually abated and, by the time they reached Flaxpool Lane, there was a beautiful stillness in the night air, save for a distant dog barking or rustlings in the nearby hedgerow. They stood at the edge of the pool, embraced and kissed tenderly at first. Sarah's warm breath on his face intoxicated him.

He pressed his open mouth on hers and slid his hands from her waist down along her firm hips. Sarah instinctively stepped back and found herself sinking into the ooze at the edge of the flaxpool. She gave a little shriek and leaped out onto the lane.

"Gordon Johnston, are you trying to drown me?" She laughed nervously.

"With kisses, yes."

"Well not tonight! If I don't get home soon Bridie will come looking for me."

"Don't mind Bridie! She'll have her hands full with Jimmy McCabe."

"Jimmy'll get short shrift – padded crossbar and all! Goodnight, love." She gave him a quick peck on the cheek and skipped down the lane before he could embrace her again.

"When will I see you again?" he called.

"You'd never know," she giggled coyly, "and keep your voice down. You'll have Canon McKenna out with his ashplant!"

"Wouldn't I just love to meet him face to face?" he shouted even louder.

"Shush! I – I might take a walk through Culvinney – Sunday evening – after tea – if it's fine. 'Bye!"

He blew her a kiss as he mounted his bike. As she turned into the gate of the McKevitt farm, she could still hear his whistling echo across the dewy fields. Bridie was stooped over a glowing fire, stirring a simmering pot. "You took the long road home . . . " Her voice was detached. "Will you take a bowl of porridge?" she added quickly.

"Of course she will. Courtin's hungry work." Sarah was initially startled by the voice that came from the wicker chair by the dresser but its unmistakable soft burr caused her to turn and leap in delight.

"James!"

Her brother rose awkwardly and came into the light to greet her with a warm embrace. He was the only brother who would greet her thus. If it were Pete or Hughie, they would mumble an awkward "hello".

"How – where on earth did you come from?" Sarah's eyes danced with delight and surprise.

"Did you not meet my chauffeur on the road – or were you too busy at other things?" He wagged a finger at her. "I got the train to Newry and, as luck would have it, when I came out of the station wasn't I nearly knocked down by Podgy Kernan's pony and trap – his aunt Daisy died in the hospital yesterday."

"I know – I heard – but why – you should have let us know – "

Bridie plopped two bowls on the table. "Eat it while it's hot, let ye."

James nodded at Sarah and limped over to the table. Sarah sat hurriedly opposite him. "Now tell me all – no, eat up first! You've had a long journey!" She smiled at him through the steam that rose from the bowl that Bridie had slid in front of her.

James. Her eldest brother. There had always been a special bond between the eldest and youngest of the McKevitt family. When their father died in the influenza epidemic of 1918, James became a father to the "wee bairn" as he called Sarah. And then two years later tragedy had struck him down. He was coming home from the creamery with the horse and cart when gunfire from a trigger-happy load of Black and Tans caused the horse to bolt. James was pitched forward between horse and cart and dragged for almost a mile down the road. His right leg was mangled. There were fears that he would lose it. He didn't, but botched surgery ensured that he would drag a shortened leg around in pain for the rest of his life. His farming days were over. He would never succeed to the farm as was his right. He assigned the farm to Hughie and Pete and for two years he suffered and brooded. It was Sarah who noticed how good he was with his hands, especially in soling a pair of boots. Finally Canon McKenna, through a cousin in Dublin, had found James a place in the capital city where he would serve his time at the leather trade – cobbling and saddlery. She would always be grateful to the Canon for that. It was the redemption of James. Though desperately lonely and painfully conscious of his disability, he worked hard and learned his trade well. He wrote regularly to Sarah. She cherished his humorous observations of city life and the society for which he worked. Often when she felt lonely herself, she would take down the tin biscuit box where she kept all his letters, pick one or two out at random and read and re-read them.

And now, out of the blue, here he was sitting across the table from her, eating his porridge as if he had never been away.

"Have you lost your appetite, Sarah McKevitt?" Bridie asked curtly.

"It must be love," James winked at his baby sister.

"Och aye," Bridie sighed. "Well, love never fed man nor beast!"

"Did Jimmy McCabe tell you that?" Sarah was emboldened by her brother's presence to score a point off Bridie who was clearly embarrassed.

"Well the cheek of you! Anyway, James has some news for you!" Bridie was clearly relieved to change the subject.

"Tell me! Tell me!" Sarah pushed her bowl aside.

"Well, I think I've learned all I'm going to learn in the big smoke – "

"And?"

"And I'm home for good! I'm going to start up on my own!"

"Oh James! Where? Where are you going to work?"

"Well . . . " He looked anxiously at Bridie.

"It means you're going to have to vacate the loft and sleep in my room, missy." Bridie was in control again.

Vacate the loft? She would readily sleep in the barn if it meant having her brother home again.

"Oh James!" She jumped up, sending her chair tottering across the flagged floor, and rushed around the table to embrace her brother unashamedly. It was the happiest moment she had known for years and she was going to savour it. What had begun as a most unpromising day had ended in bliss.

"Oh well," Bridie said resignedly, picking up Sarah's bowl, "I know a clatter of ducks who appreciate my porridge." There was a faint trace of a smile on her face.

CHAPTER THREE

SARAH TOOK HER CARDIGAN FROM THE HOOK ON THE BACK OF THE KITCHEN door and turned to Bridie. "I'm just going for a wee dander down the road." It sounded casual enough to include an unspoken invitation, but to her relief Bridie declined the offer. "I've a bit of sewing to do." It was a very definite refusal, something which puzzled Sarah until she met Jimmy McCabe halfway down the lane.

"I hear James is home from the big smoke," he said, dismounting from his bicycle. "We'll get a decent game of twenty-five at last!"

"I'd watch the city slicker if I was you, Jimmy. He'll take your last few pence!"

"He can try. Are you off for a *céilí?*"

"Just a stroll."

"Oh aye. Mind yourself crossing the border."

"You mind yourself. Bridie's got the darning needles out!"

They laughed knowingly at each other and went their separate ways.

The sound of excited calls and cheering from Cullyboe football field carried on the slight summer breeze but when Sarah reached the end of Flaxpool Lane she turned left, away from Cullyboe. She walked half a mile up the winding road before taking a sharp right down a grassy laneway, banked by high hawthorn hedges. The lane was little used now. The house it had once served was now a ruin, used for storing hay. Beyond the ruin, the lane led to a hazel wood, Culvinney, which was Sarah's destination now. Her heart was light as she sauntered down the lane, plucking at the hawthorn leaves that danced in the evening breeze. She heard the whirr of bicycle wheels behind her and turned in anticipation of greeting Gordon. To her dismay it was Pat Trainor who swerved his bike up close to her.

"Taking the air, Sarah?"

"It's a grand evening."

"'Tis that. You come down this way a lot – I've seen you."

She plucked another hawthorn leaf.

"Going to meet himself, I suppose."

"What if I am?"

"He's not your kind, Sarah – "

"And you are?"

"Well, you thought so a year ago."

"You can learn a lot in a year." The colour drained from his face at this rebuff.

"If I was – " He checked himself and lowered his voice in an apologetic tone. "I was good enough for you then, Sarah. You're better off with your own kind." He placed his hand on her shoulder. She shrank instinctively from his touch.

"Yourself and the Canon must be comparing notes," she said icily.

He caught her hand firmly. The clamminess of his grip made her feel distinctly uneasy.

"Sarah, I only want – " Both of them were startled by a shriek from the unkempt figure who dashed towards them from the ruined farmhouse. Jemmy Patchy levelled his staff at Pat Trainor and advanced towards him cautiously.

"Taking the Bulgar prisoner, sir!" He slid the words from the corner of his mouth, never taking his eyes off Pat.

Sarah smiled at the humour of the situation but principally she smiled with sheer relief. "He's one of ours, Jemmy," she whispered, only realising the irony of the words as she uttered them. Jemmy stood his ground, his blue eyes blazing at Pat.

"Ah go to hell, Jemmy, ye two ends of a fool!" Pat Trainor jerked his bike around viciously and began pedalling back up the laneway. Jemmy's eyes darted from the retreating figure to Sarah and back.

"God bless you, Jemmy." Sarah brushed her hand against the sleeve of his greatcoat. He stood to attention and saluted as Sarah hurried towards the hazel wood.

There was no sign of Gordon when she reached Culvinney.

Normally she would have enjoyed a stroll through the cooling arches of the hazel thicket, savouring its light and shade, but following her experience with Pat Trainor she was ill at ease and walked briskly through the wood. At the far end of the wood, the terrain thinned out into scrubland which rose sharply to a rock-strewn knoll. Sarah picked her way along a narrow path to the top of the knoll. Here she felt less threatened. She had a clear view across to Caraher's Hill and, within moments of her reaching the top of the knoll, her heart lifted when she saw Gordon's erect figure career down the hill on his bicycle. She hummed to herself as she wandered into a coomb on the side of the knoll. This sheltered hollow was dominated by a huge flat-topped boulder – the Mass Rock where, a century ago, the outlawed priest celebrated Mass with his people gathered on the side of this rocky knoll. Sarah loved to close her eyes and run her fingers along the top of the rock. It brought her close to previous generations.

"Is the Mass nearly over?" The familiar voice woke her from her reverie. "Or will I come back later?"

"You're not to be making a mockery of my faith, Gordon Johnston," she chided him, "and you're late!"

"Regarding the first charge – my, but we're touchy! Regarding the second, guilty – but I'm trying to get the hay cocked before this weather breaks."

"I'm not being touchy. It's just that this place – Coomataggart, the Priest's Glen – is special to all of us – "

"Us?"

"Yes us – Catholics."

"Prods keep out, in other words."

"Oh Gordon, this is silly." She put her arms around him and held him tight. He kissed the top of her head.

"Yes it is. I'm sorry." She smiled up at him. Their lips met in a warm lingering kiss.

"I nearly collided with a friend of yours at the foot of the brae." He laughed as they walked hand in hand back towards the hazel wood.

"Oh, Jemmy Patchy? Poor old Jemmy – "

"No, not Jemmy. Pat Trainor."

He felt her hand go taut as she stopped momentarily.

"He's no friend of mine, Gordon. You know that."

"Yes but he was – "

"The past is the past. I just don't want to talk about him."

He sensed the tension in her body.

"Well he's no friend of mine either." Gordon laughed uneasily. "When he didn't cut me in two with his bike, he certainly cut me in two with the look on his face."

They walked through the wood in a fragile silence. Sarah wanted desperately to rid her mind of Pat Trainor.

"Have you much of the hay to save?" she asked, almost unconcerned.

"Too much. Young Brian Caraher was to give me a hand but he never turned up."

"He probably went to the football field."

"Aye and from there to Podgy Kernan's – I suspect old Caraher put a word in his ear anyway, 'Have no truck with them Unionists,' he'd say."

"Things aren't that bad, surely?"

"You'd be surprised."

"So you need another pair of hands?"

"Have you anyone in mind?"

"How about these?" She cupped his face in her hands.

"Ach, a wee girl like you! What would you know about cocking hay?"

"You just give me a lift on your crossbar, Gordon Johnston, throw me a pitchfork and I'll learn ye about cocking hay!"

He beamed a smile of surprise down on her, kissed her forehead, grabbed his bike and motioned to her to seat herself on the crossbar as if it were a gilt carriage. They sped down the laneway, laughing as the bicycle wobbled on and off the grass verge to avoid potholes. Sarah's hair streamed into his face. He savoured its delicious perfume. The earlier fragility was gone. He began to sing, "Roamin' in the Gloamin'." Sarah leaned her head back into his chest. She knew he was happy when he sang his favourite song.

They dismounted at the foot of Caraher's Hill.

"Might as well save our energy for the hay," Gordon panted. Sarah stopped suddenly and clapped her hand to her mouth.

"Oh, I completely forgot! How could I have forgotten?"

"Forgotten what?"

"I forget to tell you James is home – for good."

"For good? But what's he going to do?"

"He's going to start a cobbler and saddlery business at home – in the loft. It means I have to move in with Bridie. But of course that won't be for long," she added shyly. "Isn't it great news?"

"Surely." He knew how fond she was of James. "I just hope he makes a go of it."

"Well, you'll all have to help him. Put the word around that the best cobbler in Ireland is in Drominarney. And certainly if you want a bridle made, or if Rose wants her boots repaired and you don't come to James, then it's all off between us!" She wagged a finger playfully at him.

They reached the hayfield and clambered over the rusty gate.

"Which reminds me," Sarah said as she surveyed the hayfield, "how is Rose?"

"No different. She has her hands full looking after Mother. It's not easy for her."

"She hasn't changed about me either, I suppose?" Sarah asked, chewing a blade of hay.

"She says little. She's – cautious."

"I know. There's a few of them in our house too. But not James!"

"Well here you are," Gordon called, tossing her a pitchfork. "Let's see how good you are with thon!"

"You'll see! I just hope you pay well!"

"Oh very well. In kind of course." He squeezed her waist as he passed.

They worked hard, talking little but humming snatches of songs to themselves, content in each other's presence and finding satisfaction in making steady progress with the work. Once Gordon stood in admiration of Sarah as, with earnest face, she deftly gathered the hay into one big forkful before tossing it onto the base of a haycock.

"Well come on! What are you standing there for?" she chided him. "Am I to be doing all the work around here?"

The fading light and the rising dew finally forced them to call it a day. Gordon leaned on his fork and shook his head in satisfaction. "There's only a half day's work left there. We did well. You're a brave wee worker, Sarah!"

"Won't it be well for the man that gets me?" she teased. She shivered slightly as the chill of the summer night assailed her sweating body.

"Here!" He draped the cardigan over her shoulders. "I don't want to lose a good worker now that I've found one. I'd better take you home."

"The Rolls Royce awaits?" she purred.

"At yon gate." He gestured towards the bike with a sweep of his hand. They both relished the cooling breeze as they freewheeled down Caraher's Hill. When they reached McKevitt's gate, Sarah slid off the crossbar and took Gordon's hand.

"You'll come in – for supper – and to meet James?"

"It's late. I should be – "

"Please?"

He shrugged his shoulders and let the bike fall against the hedge. As they entered the porch, Bridie's monotone voice within caused them to pause.

"The fifth Glorious Mystery, the Coronation of Our Blessed Lady. Our Father . . ." Gordon turned to go but Sarah tightened her grip on his hand, drew him towards her and kissed him very tenderly. The kiss lasted for the entire decade. While Bridie went through the trimmings of the rosary, the two stood in the porch making faces at each other like naughty children. A scraping of chairs on the kitchen floor signalled the end of prayers. Bridie was ladling out porridge when her sister entered.

"Good timing," Bridie said without turning around. There was heavy cynicism in her voice. "You'll take a bowl of porridge, Gordon?"

"Ehm – " Sarah squeezed his hand. "Aye, if it's not too much trouble."

Hughie and Pete muttered a hello to Gordon and set to eating their porridge. Sarah led him to the wicker chair.

"James, this is Gordon."

James rose awkwardly from his chair and the two men exchanged greetings. Bridie placed a bowl of porridge before each of them as they indulged in desultory conversation about the weather, saving the hay and life in Dublin. When Hughie went outside to check the stock and Pete slipped away to bed, James finally brought up the question of marriage.

31

"Well, when is the big day?"

"October – late October – we haven't fixed the date yet," Sarah answered, eyeing the two men. "The banns will be read out in a couple of weeks."

"I wish ye luck – and God's blessings," James said. Both Gordon and Sarah felt the sincerity of his words.

"James, I have a wee favour to ask of you."

"I suppose you want a new pair of shoes for the wedding."

"No, not that," Sarah laughed. "Hughie will sell the black bullock for me at Cullyboe Fair in August. No – I want you to give me away at the wedding."

"Is that all? It will be my honour."

Sarah kissed her brother in gratitude. An unnatural rattling of pots and bowls came from the back kitchen where Bridie was washing up.

"I really must be going," Gordon said. "It was nice meeting you, James – and thanks for the porridge, Bridie," he called out. An even louder rattle came in reply.

Gordon looked puzzled but Sarah reassured him, "That's just Bridie's way." She accompanied Gordon to the door and blew him a kiss. When she returned, Bridie was busily wiping down the oilcloth on the kitchen table.

"One decade wouldn't have killed you," she muttered.

CHAPTER FOUR

THE WEDDING WAS FINALLY ARRANGED FOR OCTOBER 27TH, THE LAST Saturday of the month. Sarah had to endure a further lecture from Canon McKenna before he grudgingly read the banns of marriage. As Sarah sat in the front of the church that Sunday morning, she felt a hundred pairs of eyes riveting the back of her head at the mention of her name. Canon McKenna made a few deliberate pauses, as if inviting objections from the congregation there and then. Outside the church a few neighbours nodded and nervously wished her good luck. Jimmy McCabe was the exception. He strode brashly up to her, put his arm about her shoulder, and proclaimed in an embarrassingly loud voice, "Good catch, Sarah! Strong farmer, fine house, bit of money in the bank – sure they're all jealous of you!" He gestured towards the curious onlookers, adding in a whisper, "What matter if he's a Prod!"

"Jimmy McCabe, you're a total scourge!" She couldn't make herself sound convincing and broke into a shy laugh. "Any chance you'll make it a double wedding?"

"Oh now, that would be rushing things totally! I'm appraising the situation, shall we say. I suppose her ladyship went to early Mass?"

"She did. You'd need to change your habits if you want to impress there, Jimmy!"

"No, no. That's not the way of it at all. You see, if she was lucky enough to get me, she could give me the breakfast in bed on her way to early Mass!" He lit a cigarette and winked broadly at Sarah as he blew out the match.

"Go out of that, you hypocrite!"

"Give her ladyship my compliments. Tell her I shall come a-calling as soon as I have the spuds sprayed!" He gave a little leap and skipped off to join his cronies at the church gate. Sarah gave him a cheery wave. It was the antics of the Jimmy McCabes of this world that preserved her sanity. She wheeled her bicycle down the path to the cemetery partly to say a prayer at her parents' grave but partly also to avoid the stares and whisperings of the worshippers of Cullyboe.

The summer months slipped by uneventfully – too uneventfully for Sarah. She would have liked if her impending marriage were discussed by her family. Bridie never broached the subject despite Jimmy McCabe's tauntings. Hughie and Pete, when they did talk, talked only of football and potato blight. Even James avoided the subject. He was slowly building up his business and almost all of his energies were channelled into that. The loft was transformed into a workshop. Hughie and Pete built a workbench which ran the length of the room. James wrote to his former employer in Dublin and, within a week, a large consignment of leather, waxen thread, nails and various tools arrived at the railway station in Culloville. James was a very meticulous worker and arranged the tools and materials of his trade neatly around the room. Sarah loved to visit the loft, if only to inhale the comforting smell of leather and to marvel at the array of tools that hung on pegs along the wall. As James gradually picked up work, the number of visitors to the house grew and brought added interest to the life of the McKevitts. Although the customers gained access to the loft by an external stone staircase at the gable end of the house, their comings and goings were acutely observed by all of the family and were often the source of livelier than usual conversation over supper.

"Hasn't Barney McQuillan aged a terror?" Bridie would say. "I didn't recognise him going up to the loft with the stoop that was on him."

"He suffers a lot with the back," James would explain.

"He suffers a lot with them six daughters of his, if ye ask me," Hughie would add with a mischievous grin, "every one of them lazier than the next – "

"And no oil paintings either!" Pete guffawed.

"Maybe one of you should take a daughter off Barney's hands," Sarah suggested.

"Not if I was paid a king's ransom," Hughie countered. "Sure they'd scald you. Never do a hand's turn about the place! Isn't that what put their poor mother in an early grave – doing everything for them?"

"As long as they wear out their boots I'm happy," James said with a quiet smile. And then they would move on to another visitor.

As the weeks slipped by Sarah grew more anxious. It's almost as if I'm not getting married at all, she thought. I suppose they're hoping I'll change my mind and the wedding won't really happen. She decided to speak her mind to the one person who would listen to her.

"I can understand your anxiety," James said as he pulled a thread through his teeth before running it through the eye of an awl. He cradled a horse's collar in his lap and began stitching it together. "But it's difficult for them too. For us. This is a very small community. If it was in Dublin it would be different. But here . . . And then there's this border business."

"But you've met Gordon. He hasn't got two heads or anything – "

"Of course he hasn't. He's a fine fellow. But he's still an outsider. He's from the other side. And you know how suspicious we are in this country of anyone from the other side. We've been killing one another for the past few years because people came from 'the other side'." Sarah, who had been idly fingering a leather thong, suddenly found herself wrenching at it with both hands in frustration. It snapped.

"I'm sorry." She smiled weakly in embarrassment. "I'd better go before I do any more damage! There's one more thing, James. Gordon wants us to visit his home before the wedding. There's only himself and his sister Rose – and his mother. She's confined to bed all the time. Will you speak to the others?"

"I'll speak – but I don't know if anyone will listen."

James's doubts were realised on the appointed Sunday when Hughie became very anxious about a sick cow and couldn't leave her. Pete had to go to Castleblayney with the Cullyboe junior football team. Bridie obviously only acceded to James's pleas under protest and was in a sullen mood as she harnessed the pony and trap. There was little conversation among the three of them on the three-mile journey. An occasional question was asked about the ownership of a particular field or a comment was passed on the state of a crop. As the trap swayed and bounced on the uneven road, Sarah grew more apprehensive about the visit.

Sloe Hill was a solid, if featureless, two-storey house set in the shelter of a yew copse. Its stone-dashed appearance was unwelcoming. When James pulled up the trap on the gravelled front, a black and white collie appeared suddenly from the side of the house and barked loudly at the visitors. Gordon opened the front door and gave a sharp command to the dog, who slunk back to the side entrance, giving the occasional subdued bark.

"Don't mind him. He's all bark." He smiled, extending his hand in welcome. He was dressed in his Sunday best – a bright navy-blue striped suit with a high-collared blue shirt and tie. Although he looked distinctly uncomfortable, Sarah was proud of his appearance.

The McKevitts were led through the entrance hall to the parlour, a room which cried out for redecoration. Its windows did not catch the afternoon sun and its gloominess was accentuated by the dark brown embossed wallpaper, an even darker suite of furniture and faded green brocade curtains. A huge oak sideboard ran almost the length of one wall. Gordon fussed as he seated his guests and made comments about the weather.

"Would anyone like a drink?" he asked nervously as he prised the door of the sideboard open. "Sweet sherry or some cordial? No spirits I'm afraid," he added looking at James. "It's a rule of the house on Sundays."

James waved his hand. "We're grand, we're grand," he assured Gordon. Bridie declined politely with a "No, thank you."

"I'll take some cordial, love," Sarah said springing to his side in support. "Let's see what you have!"

As both of them peered into the sideboard, she whispered to him, "Relax, Gordon. They won't eat you." She took a glass of raspberry cordial and resumed her seat.

"Rose will be along with the tea shortly anyway. She's up with Mother at the minute."

"How is your mother?" James asked.

"Ah sure – the same. Never changes. In a world of her own. A world of her own." He looked out the window into the distance.

"That's her there – as a young woman," Sarah pointed to a portrait of a solemn-faced woman which hung over the mantelpiece.

"A handsome woman," James commented.

"You're like her," Bridie added in a more mellow voice. Sarah appreciated that.

"You'll call up to see her before you go," Gordon insisted. "Rose will – "

As if on cue, Rose entered the room with a fixed smile, smoothing the sleeves of her cardigan.

"Ye're very welcome," she said curtly. "I'm sorry – I had to tend to Mother."

Sarah introduced her brother and sister to Rose who then excused herself and returned pushing a tea trolley which had obviously been prepared well in advance of the visit. The tinkle of fine china punctuated the dragging silences in the conversation. A grandfather clock chimed in a distant room. Sarah decided to bring up the topic they were all avoiding.

"About the wedding," she began nervously, "it will be a small affair – just ourselves. Unless some more of your family would like . . . "

"Uncle Trevor can't come," Gordon said quietly, "can't take time off. He's training-in new staff in the shop. He sent us a present – of a chiming clock. It's in the dining-room."

"That's nice." Sarah smiled weakly.

If there was one thing Sloe Hill did not need, it was another clock. Life seemed to tick away at a different rhythm in every room.

"And Aunt Madge will have to stay with Mother while I'm away," Rose added stirring her tea very intently.

"I understand," Sarah said, "the wedding is early – at nine, in Cullyboe."

"The Canon'll not keep us long." James's attempt at humour only achieved another awkward silence.

"Would anyone like more tea?" Rose inquired. There was a general shaking of heads and waving of hands.

"Your wee fairy-cakes are lovely," Bridie said graciously.

"And that's a compliment coming from Bridie!" Sarah added quickly. "You should see the wedding cake she's baked."

The grandfather clock chimed again.

"I think it's time we made a move," James said. "I have a man coming to collect a collar." They all rose as one.

"You'll see Mother before you go?" Gordon suggested. Rose flashed an anxious look at her brother. "She'll not know you, of course," he added, "but just to say hello."

The bedroom was narrow and sparsely furnished. A tiny white-haired head propped against pillows was all they could discern of Mrs Johnston in the darkened room. Only a chink of sunlight penetrated the heavy drapes.

"This is Sarah, Mother," Rose said, smoothing the counterpane, "Gordon's intended."

Sarah felt the doubt in her voice.

"And Bridie and James. They just came to say hello." The little group stood awkwardly in the doorway.

"Well isn't it the brave cold day?" the little voice came in reply. "Harold said he couldn't get the horse up Caraher's Brae with the frost and the ice. Poor wee Gordon'll be petrified in school. Master Carson never lets them near the fire . . . "

"Aye indeed, Mother," Rose agreed and motioned to the group to withdraw.

"She's like that all the time," Gordon explained embarrassedly. "She's very well in herself but her mind's away."

The McKevitts thanked their hosts and took their leave. The chill of the October evening pinched at Sarah's face as the trap sped down Caraher's Hill. Again there was silence on the journey home. It was only when they turned down Flaxpool Lane that Bridie commented, "It'll not be easy – sharing a house with two other women."

As ever, James tried to ease the situation. "Sure it could be no worse than sharing one with the four of us!" Sarah smiled in appreciation of his comment but she knew that Bridie had a point.

CHAPTER FIVE

IN HIS SERMON ON THE SUNDAY BEFORE THE WEDDING, CANON MCKENNA made what seemed to Sarah to be his last pitch against her marriage to Gordon Johnston. It began with words which caused a good deal of nudging among the men who crowded into the porch of the church and knelt, one-kneed, on their caps on the red-tiled floor.

"My dear people," he began, "we live in changing times. Changing and very dangerous times. The faith and the strong moral fibre that previous generations defended proudly, even unto death, that faith is under attack. It is not an open attack. There are no marching battalions on the horizon. No, my dear people. Would that the enemy would show itself on the open plain of the battlefield. Then indeed we could march together to meet it and crush it without mercy." His fist thudded into the red padded cushion that ran along the top of the pulpit rail.

"No. This is a much more dangerous and insidious enemy. An enemy that weaves and slithers its way into the heart of our society just like the serpent in the Garden of Eden. An enemy that takes many forms. I think especially of the evils of the dance-hall which our bishops rightly spoke out against earlier this year. I do not speak of the schoolhouse dance which is properly supervised and supportive of our own music and culture. I speak of the modern dancing-halls which have begun to proliferate around this country. Indeed we have one not five miles from where we are gathered today." His voice suddenly rose in pitch and vehemence.

"Importations from the vilest dens of London, Paris and New York," he thundered, the fist ramming rapidly into the cushion. "Direct incitements to evil thoughts, evil desires and the grossest acts of impurity. Not my words but those of our bishops. What else could these places be, with their foreign dancing and their jazz music! Can't you imagine the anguish of the souls of the faithful departed – the souls of those who gathered at the Mass Rock in Coomataggart when the priests were hounded from their churches – can't you imagine their anguish and despair to see the faith they clung to 'in spite of dungeon, fire and sword' being cast aside by our young people in favour of the fleeting pleasures of the body!" His hands gripped the cushion fiercely and his spittle rained down in a fine spray on the altar boys seated beneath him.

"We must be on our guard, my dear people, for the enemy stalks our land. Parents, be vigilant over your children. Brothers, protect your sisters. Young women, pray that you do not succumb to temptation." His voice dropped again. "Of course the enemy is not just in the dancing-hall. Oh no. It is all around us. In the books we read, the pictures we look at. In the very thinking that gnaws at the fabric of our society, like the rat that burrows into the potato-pit and destroys the fruit of our hard work. There are those who consider it fashionable to ignore the family rosary – worse still the frequent reception of the sacraments. It is considered by some to be quite in order to marry outside one's faith." He paused. Sarah closed her eyes and winced. The voice boomed again.

"I say no! No and no and no again! No to the easy road of loose and liberal thinking. Our Lord and Saviour Jesus Christ never said salvation would be easy. It would be easier for a camel to pass through the eye of a needle . . . So my dear people. Take the hard road. Stand up to the enemy. Stand up for your faith. Be watchful of yourself and of your neighbour. Follow the example of our parents and the generations who have gone before us. Blessed are the meek for theirs is the kingdom of heaven. Amen. We'll stand now and to show our resolution, we'll sing a verse of 'Faith of our Fathers' . . ."

Canon McKenna's booming voice rang in Sarah's head all day. No and no and no again . . . By late afternoon she could no longer take the confinement of the house. She strode briskly down Flaxpool Lane and as she turned left towards Drominarney she almost collided with

40

Jemmy Patchy. He stood stiffly to attention and saluted. "Permission to visit the village, sir? On leave, sir."

"Permission granted, Jemmy." Sarah smiled. "Enjoy your leave!" He clicked the broken heels of his boots together, turned and marched off in the opposite direction singing loudly:

"Pack up your troubles in an old kit bag
And smile, smile, smile . . . "

Sarah smiled and continued her walk with a somewhat lighter step. Jemmy never failed to lift her spirits. She had intended to pay Gordon a surprise visit but the increasing chill of the evening and the oncoming dusk caused her to change her mind. She turned down the lane to the hazel wood. The autumn wind ruffled the growing carpet of leaves in the wood. Sarah trailed her feet through the leaves and paused to lean in thought against a tree. She cradled the bole of the tree in her hand.

Suddenly her hand was clamped to the tree by another hand, larger, fleshy and clammy. Sarah felt an icy spasm shoot down her spine. She knew who it was before the voice spoke from an uncomfortably close distance behind her.

"I hope you were listening closely to that sermon."

Sarah felt the warmth of Pat Trainor's breath on her neck. She struggled to free her hand but he had locked it firmly in his grip.

"You're hurting me," she whispered.

He released his grip but immediately restored it with his other hand so that he could swing himself around the tree to face her. She tried to avert her face but his face followed hers.

"Why are you doing this? Why are you – haunting me like this?" she pleaded.

"Because it's not too late, Sarah." His breath was now on her face. "It's not too late to back out. To back out of a mistake!"

"It's not a mistake!" She shook her head violently. "It's my choice! It's my life!" She forced herself to look brazenly into his eyes.

"It's not right! You heard the Canon – "

"Did you write the sermon for him?"

"You're only fooling yourself, Sarah." There was a harder edge to his voice. She felt her trapped fingers go numb.

41

"You're hurting me – please let me go." Her voice broke into a sob. He took a step closer to her, peering into her eyes.

"You know I wouldn't hurt you, Sarah," he whispered, "you know I wouldn't hurt you!"

The last few days before the wedding were a blur of preparation. Hugh gave Sarah the proceeds of the sale of the black bullock that had been "hers" since it was a calf. She and Bridie took the train to Dundalk for a day's shopping. They completed their business quickly – Sarah knew exactly what she wanted and her budget was limited.

"Tell you what," Bridie said as they stepped out of McConville's Drapery Store, "let's have lunch in the Railway Hotel. My treat!"

"But what about the men?" Sarah began.

"The men will survive without us. James is a good cook. He can heat up a stew at least," said Bridie briskly.

Lunch in a hotel was a rare treat for both of them. They initially felt conspicuous, almost apologetic for their presence, but the arrival of a raucous and rather drunk farmer soon deflected attention from them.

"You're very quiet," Bridie said as soup was served.

"Haven't much option with that fellow ranting away over there!"

"I don't mean that. Over the past few days – you haven't said much."

"I'll have the chicken – and spuds and turnips," the farmer bellowed. Sarah smiled nervously.

"You're not having doubts?" Bridie continued.

"No. I'm not having doubts. I'm getting married on Saturday," Sarah retorted firmly.

"And there's nothing – troubling you?"

"No. Why should there be?" Sarah was growing quite edgy.

"Nothing." A fragile silence hung over the main course.

"You were very restless in your sleep these past few nights," Bridie said at last. "You kept calling –"

"Who – what – was I calling?" Sarah gripped the tablecloth anxiously.

"Well, Jemmy Patchy of all people!"

Sarah laughed nervously. "My secret is out – Jemmy and I are planning to elope! What about yourself and Jimmy McCabe?" she added quickly.

"What about us?"

"Well, are ye going to give us a day out?"

"Me and Jimmy McCabe?" Bridie spluttered through a mouthful of ice-cream. "You're not serious?"

"I am – and so is Jimmy!"

"Jimmy McCabe is a fool. Oh, a likeable rogue, but a fool nonetheless."

"You could do a lot worse!"

"I wonder," Bridie mused.

There was another long silence which was finally broken by the loud farmer. "Where did yis get the chicken? I seen more mate on a sparrow!"

The sisters exchanged smiles.

"I think it's time we left," Bridie whispered.

There were but a few curious onlookers in the village of Cullyboe when the McKevitt family arrived in two traps shortly before nine o'clock on the Saturday morning. It was a cold, blustery morning. The leaves whirled in circles under the legs of the two horses, causing them to stomp restlessly as Hughie and Pete tied their reins to the church railings.

"Cullyboe should have the brass band out to meet us," James laughed as he helped Sarah from the trap. "Aren't we the fine prosperous-looking family all in our best suits and with two horses and traps!"

Sarah looked around at her sister and brothers. She hadn't seen them all dressed so well since their mother's funeral. Yes, they were a handsome family.

"This cursed pony of Podgy Kernan's is a very flighty animal," Pete snapped. "I'll not be surprised if she takes off home while we're inside."

They hurried into the church. They knelt before the altar but Canon McKenna hastily beckoned them into the sacristy. Gordon was waiting for her there. He was accompanied by Rose and a burly man with thickset features whom Sarah had not met before.

"Good luck to you both," James whispered.

"This is Winston Pollock," Gordon introduced the burly man, "he's my best man." Winston Pollock nodded towards each of the McKevitts.

The wedding ceremony was performed in a cold perfunctory manner. Canon McKenna seemed intent on getting through the wedding in the quickest possible time and with the least personal involvement. They signed the register in silence. The Canon closed the register and put it away in a tall press, which he locked carefully. As he did this he spoke in a detached voice, "I regret I am unable to attend the wedding breakfast. I have Mass at eleven and a number of sick calls thereafter. I wish you both God's blessing."

"Thank you, Canon," Sarah whispered. He was surprised when Gordon offered his hand in gratitude and took it embarrassedly.

Outside the church Sarah was surprised to find a shining black Baby Ford awaiting them. "It's Winston's," Gordon explained, "now you see why I chose him as best man!" He kissed her very gently. "You look very pretty," he whispered. "I'm real proud." She clenched his hand tightly. They climbed into the back of the car. Bridie slid nervously into the front. Just before Winston Pollock got in, a slight figure wearing a long tweed coat and a headscarf appeared at the door. It was Aggie Trainor, Pat's sister.

"I wish ye health and happiness. I really do," she said in a choking voice. Before Sarah could say a word in reply, she was gone.

They drove back to Sarah's house for the wedding breakfast. Bridie had worked hard the day before to ensure that it would be an enjoyable meal. They ate well in a restrained atmosphere. Gordon attempted conversation with Hughie and Pete about the price of cattle but they had to excuse themselves to tend to farm duties. A few neighbours called to wish the couple well. Matt McLarnon brought his melodeon and played a few tunes. Jimmy McCabe sang "Slieve Gallion Braes" and wisecracked his way through the morning. Rose Johnston maintained a sedate presence throughout the celebration but Winston Pollock was totally ill at ease, excusing himself every thirty minutes, ostensibly to attend to some mechanical problem with the car.

At two o'clock Gordon announced that they would have to leave in order to catch the Belfast train.

"The Belfast train?" Sarah queried.

"Aye. We'll spend a short honeymoon there. Only a few days, mind. I thought you'd be pleased."

"I am! I am!" Sarah cried with delight. She kissed her husband to the applause of all present.

"Just take what you need in the valise. We can send the rest over to Sloe Hill while you're away." Bridie's voice trailed off over the last few words.

Sarah put her arms around her sister and hugged her. "Thank you – for everything," she whispered, the tears welling in her eyes.

"Go on – or you'll miss your train," Bridie replied, trying very hard to sound indifferent. Hughie and Pete came in from the farm to say goodbye. James pressed a small package into Sarah's hand.

"You can open it later. Just a little memento."

She held her brother tight as the tears rolled down her cheeks. Winston Pollock revved the engine of the Baby Ford. Sarah and Gordon climbed into the back and Rose sat in the front. She had to return home to tend to her mother. As the car swung out of the yard, Sarah waved furiously to the little group assembled at the gable end, by the steps to James's loft. It was only now that she realised she was leaving her home for good. A great emptiness gnawed within her. She opened the package James had given her. It was a leather purse, beautifully crafted, with the name SARAH delicately carved under the flap. Inside was a crisp new five pound note. She pressed the purse to her damp cheek. Gordon took her other hand.

As the car slowed to turn right at Flaxpool Lane, an unkempt figure suddenly appeared from behind a clump of bushes. Jemmy Patchy stood stiffly to attention, clutching his staff with his left arm while his right arm rose in salute.

CHAPTER SIX

BRIDIE McKEVITT SMILED AS SHE RAN THE DRIPPING CLOTHES THROUGH THE wringer. A familiar jaunty whistle came over the March air as she emptied the remaining soapy water over the flags in the haggard. She paused to watch the water find its way into the crevices between the flags. "Yon fellow surely likes to announce his arrival," she thought before staggering across the haggard to the orchard with a basinful of washing.

"Begod if I'd known you were washing I would have brought you a bag of me own," Jimmy McCabe chirped as he slid off his bike and vaulted the gate into the orchard.

"I have enough to be going on with here," Bridie muttered through a mouthful of clothes-pegs, "with three men in my life."

"Come on, Bridie, you know there's only one man in your life!"

"Hope I know him when I meet him."

"Oh, we're in one of those moods, are we?" Jimmy said. He took up the basin and followed Bridie along the clothes-line between the two apple trees.

"No moods, just facts."

Jimmy felt the rebuff keenly. "Any news of the newly-weds?" he asked, sensing the need for a quick change of subject.

"Yes – there's a baby on the way!" Bridie's voice brightened considerably. "Due in July."

"Well that's a bit of good news to be sure. Due in July?" Jimmy did a quick mental calculation. "Begor, Gordon wasted no time there.

They say the Prods always – " Bridie froze him with a look as she took the empty basin from him.

"Is that bit and bridle for yourself?" she enquired icily, eyeing the harness that was draped over Jimmy's arm.

He laughed nervously. "No. Wouldn't fit! It needs a bit of stitching. Is the saddler about?"

"Hard at work in the loft as ever." They walked back through the haggard.

"Any news from your part of the world?" Bridie asked.

"Ah sure, not much. The oil wells are pumpin' away and the old man has just bought out the bank – "

"Ah Jimmy. Be serious for once!"

"Well I heard there was a fellow fished out of Lough Muckno with a bullet in his head. The war isn't over for some, it seems."

"Indeed. Podgy Kernan was saying we should have fought on. That we'd rue the day the Border came – "

"Hah! That tub of lard! Whenever there was any bit of fighting, the same Podgy was under the counter hidin' the takings. All he's rueing is the loss of trade!" There was a silence, broken only by the tapping of James's hammer above them.

"Anyway how come you were talkin' to Podgy?"

"He delivered a sack of flour to me. Don't tell me you're jealous?" Bridie laughed.

"Jealous?" Jimmy spluttered as he ascended the stone stairs. "Jealous?" He fought hard for something to say. He reached the top step and stared across at Lake Muckno in the distance. He turned and reached for the door handle.

"I'll tell you something. You'd much prefer to be doin' my washing than Podgy Kernan's!" He disappeared into the loft before Bridie had time to react. She smiled as she swung the basin by her side on her way back to the kitchen.

Three chiming clocks argued the exact time as Gordon Johnston came into the kitchen for his tea.

"There's a brave cold wind out there," he said splaying his outstretched fingers before the fire. "We might have snow yet."

"Don't say that," Rose said as she gently withdrew a cake of soda bread from the oven, "Winston's taking me to Newry tomorrow in the car. I don't want us to be slithering off the road!"

"Tomorrow?" Sarah slid Gordon's plate noisily in front of him.

"Yes. He has some business to attend to. You'll manage on your own, surely?" Rose's reply wavered between question and statement.

"But it's – St Patrick's Day. I have to go to Mass – and I thought I might visit home for a wee while."

Rose slammed the oven door shut.

"It's an ordinary Friday in this state. Nothing special goes on here. And I need to get to Newry."

Gordon watched the two women nervously.

"What time do you want to leave for Newry?"

"About twelve. I thought we'd make a day of it."

"Then there's no great problem. I'll leave Sarah to Mass and one of the boys can leave her home in the afternoon."

"And Mother?"

"I'll look after Mother till Sarah gets back."

Rose bustled about before bringing a tray up to her mother. Sarah leaned across the table and squeezed her husband's hand. He gave a genial shrug of the shoulders and continued eating with the other hand.

Sarah squeezed her way into the pew between a delighted James and Bridie.

"Is there an extra spud in the pot for me?" she whispered.

"Didn't I get Patsy Hughes to kill the pig specially for you?" James answered. Sarah stifled a giggle.

"Shush you two and have some respect," Bridie muttered in a less than convincing voice, as Canon McKenna ascended the steps to the altar.

"I won't detain you long, my dear people. I know this is a day of celebration – "

"He's away to a football match in Dublin," James whispered. Bridie glowered at her brother.

" – but we must remember who and what it is we celebrate. We give thanks for our great national apostle, the slave boy Patrick and for the most precious of all gifts he brought us – the gift of faith. Tempered by dungeon, fire and sword, that faith has become

48

engrained in our people over fifteen centuries. And, while we have occasionally lost a branch here and there, the tree of faith is deeply rooted and stands up to the strongest of storms. But we must never relax, my dear people, and think the storm is past. As with the tree of knowledge in the Garden of Eden, the serpent is ever present to tempt us, to weaken our resolve. Today more than ever that resolve must be greater. There is evil all about us – most of it coming from across the Atlantic in the form of books and music – and the latest scourge, the motion picture. So I say to you on this great feast day – be strong in your faith, the faith of our fathers. Be true to Mass and the sacraments – and I think it is fitting that on this day we send our congratulations and our best wishes to one of our own, Pat Trainor, who this week departs for England to study for the priesthood; Pat will ultimately bring the faith we proudly profess to the heathen people of Africa . . . "

The rest of Mass was a blur to Sarah. Pat Trainor – a missionary priest? Father Pat Trainor . . . She sat anxiously in the trap, avoiding the gaze of the people who filtered out of the church. It seemed as if James was conversing with half the congregation in the church yard. There was a long silence on the journey home before Bridie eventually spoke.

"Isn't that great news about Pat Trainor? I never would have thought there was a priest in him." Sarah tightened the grip on the collar of her coat as they drove into the biting wind.

"I suppose he never got over losing you, Sarah," Bridie added. Sarah looked away across the fields to the hazel wood at Culvinney.

"The heathens of Africa are welcome to him," she said quietly.

James slapped the reins across the horse's rump and cried "Hup!" as they turned down Flaxpool Lane.

There was no further mention of Pat Trainor that day and Sarah's upset was eased by being in the heart of her family once again.

"I wasn't joking about the pig," James chided her as Bridie served up a delicious bacon and cabbage dinner.

"But you didn't know I was coming – "

"Neither did the pig!" Pete mumbled through a mouthful of potato.

"Let's just say I hoped," James said.

"That was grand, Bridie." Pete jumped up from the table while the others were still enjoying the meal. "Keep me a bit of the apple tart. I'm off. I'll see you, Sarah."

"The football again," Sarah mocked. "Did ye win a match yet?"

"I'll have you know we are in the semi-final of the Feis Cup," Pete announced with pride.

"Two walk-overs and a flukey win over Emyvale," Hughie scoffed.

"Beats courtin' Josie McLarnon any day!" Pete grabbed his cap from the top of the dresser and dodged a flying potato skin as he skipped out the door.

"Well honest to God," Bridie snapped. "It's about time the two of you grew up. Yis are like a pair of five-year-olds!"

"What's this?" Sarah asked in genuine surprise. "Nobody told me there was romance in the air."

"That's because there isn't any. God, you can't bid the time of day to anyone but some people would have you up at the altar straight away." Hughie's mood had become surly and he rose to go also.

"Sit down there and act your age," Bridie commanded. Sarah glanced anxiously at James who reassured her with a playful wink. The conversation eased into small talk of who had been at Mass that morning. Finally Hughie excused himself.

"Talk never planted spuds and Father, God be good to him, always had the early ones in by Patrick's Day. I'm going out to open a few drills before the man starts turning in his grave!"

James read the *Democrat* by the fire while the sisters washed up.

"Any sign of Jimmy these days?" Sarah asked.

"Jimmy who?"

"Ah, Bridie!"

"Oh, him. He comes around all right. He was with James the other day."

"Aye," James said, his eyes fixed intently on the paper. "He's talking of going to America."

"America?" Bridie let a plate slip back into the basin of water. "He never said anything – " She paused and began scrubbing the plate furiously.

"Well he's in an awkward position," James said, turning the page. "The old man will hold the reins there till he dies and when he does Owney will surely take over. What have they anyway but thirty acres of rushes?"

"Does Jimmy have any prospects in America?" Sarah asked, not daring to look at her sister.

"He has an uncle working on the railroads."

"He'd be a loss around here," Sarah sighed.

"What loss?" Bridie snapped.

"Well, he's always good for a laugh, if nothing else – "

"That's exactly it!" Bridie snatched up the basin and swept past Sarah to throw its contents down the haggard shore. "You can't live on laughs. Someday he'll realise that."

James drove Sarah back to Sloe Hill as threatening rain clouds banked up in the sky over Cullyboe. Sarah gave a sigh as they reached Caraher's Brae.

"I don't know. There seems to be a lot of edginess in the McKevitt family these days."

"Things happen. Things change," James said quietly. "What about you? We never had a chance to – "

"I'm all right. I'll be all right."

"I was pleased to hear about the baby – "

"Yes." The trap bounced awkwardly as the horse strained to pull it up the steep pot-holed hill. Suddenly Sarah called out – "Stop, James please." A familiar figure emerged from a gateway Jemmy Patchy stood to attention.

"Awaiting intelligence from the southern front, sir. Nothing to report yet, sir."

"Well I have something to report, Jemmy. I live up in Sloe Hill now and you never come near me. Call up some day for a bit of dinner."

"I will relay that intelligence, sir." Jemmy saluted, turned and marched briskly down the hill.

"God bless the poor devil," James said, urging on the horse once more. "He's not in this world at all."

"He's in a better one," Sarah replied as she watched the ragged figure shamble into the growing gloom.

Gordon took both her hands in welcome. "Did you have a nice time?" he asked.

"It was lovely. Just to see them all again – even if they argue more than ever! And you?"

"I've been busy," Gordon sighed. "It's as if Mother senses that Rose is away. If she called me once, she called me twenty times."

Sarah put her gloves in the hallstand drawer, kissed her husband on the cheek and ascended the stairs. "You go on outside. I'll take over now," she whispered. She paused momentarily before entering the bedroom.

The tiny figure was propped up against a large bolster. Her voice belied her frailty.

"Have you got them eggs done for Father's tea? Soft-boiled, he likes them, And nobody wound the clock yet . . . "

As if on cue, a discordant chorus of chimes welled up from downstairs.

CHAPTER SEVEN

*T*HE HEAT IN THE LITTLE CHAPEL WAS INTENSE. SARAH GENTLY SWAYED HER *baby to and fro in an attempt to give him air. The press of bodies behind her grew by the moment as the enormous crowd swelled into the tiny wooden building. Sarah turned anxiously to Gordon but he was deeply engaged in conversation with a group of people behind her. The infant was growing restless. At last the priest entered through a narrow opening at the top of the room.*

Father Pat Trainor was dressed in a dazzling white soutane, the collar of which bit into his turkey-red neck. He mopped the rivulets of sweat from his forehead and waved the white handkerchief with a flourish to silence the restless crowd.

"The first baptism will be that of Patrick Oliver Johnston," he announced.

Sarah had to tug at Gordon's sleeve before stepping forward with her baby. She smiled nervously at the priest. He directed all his attention at the squirming infant, addressing him imperiously. "Patrick Oliver Johnston, I bring you the gift of the one true faith through the cleansing waters and oils of baptism, but first I ask you to renounce the heathen beliefs of your father —"

Sarah stared disbelievingly at Father Trainor. He bent over the child. Sarah found herself looking down on the priest's pink bald head as he spoke again to the infant.

"Do you renounce your heathen father?" he bellowed. Sarah snatched the child away from the priest and looked desperately to

Gordon for help. Gordon's ebony face smiled and nodded above her.

"No!" she screamed. "He's not a heathen, he's not a heathen . . . "

The hand that rested on her shoulder pressed the clammy nightdress against her skin. Sarah felt very cold. She turned, terrified, to face the owner of that hand. Gordon was sitting upright beside her, his smiling face showing both concern and reassurance.

"It's all right now. It's over. Would you like a cup of tea?" Sarah nodded, relief coursing through her body. Gordon drew the eiderdown up to her shoulders.

"Keep yourself warm. I won't be long." Sarah lay back on the pillow and clutched the eiderdown about her shuddering body.

Bridie McKevitt scattered the food scraps among the hens who came trotting from different parts of the orchard. She watched in some amusement as the hens pecked and tugged at a crust of bread. Invariably they turned on each other and fought over smaller pieces of bread.

"God bless the work!" A voice disturbed her reverie. It was Owen Brady, the postman.

"You gave me a start." Bridie laughed. "And you're such a stranger in these parts of late, I hardly knew the voice!"

"Well now, I can't help it if your admirers won't write to you more often, but maybe this is a start," he quipped, handing her a letter addressed to herself. The writing was uneven and not immediately identifiable. Bridie wiped her hands on her apron and slid the letter into the pocket.

"Any news?" she asked.

"Och, not much, except them curse-of-God Unionists cleaned up in the elections in the North. I hope they don't move the Border to bring us in under that crowd!!"

"They never would, surely?"

"You'd never know, there's a big Commission sittin' on it. I wouldn't trust Cosgrave at all. He'll probably give them a few more counties. If he does, I can tell you the guns will come out again. De Valera was right. They destroyed the Republic, them Free Staters! Good luck now!"

He was on his bike and away before Bridie could reply.

She filled the kettle and hung it on the crook over the fine, before opening the letter. The writing still puzzled her. The address puzzled her even more.

Queenstown
May 1st 1926

Dear Bridie,

Now there's a surprise for you. A letter from James McCabe Esq. If I was you I'd frame this. By the time you read this I'll be on the high sea to Amerikay. Or knowing my luck I'll probably be at the bottom of the sea. One way or another I'll not be bothering you for a while.

I'm going to give it a try in the States. Uncle Hugh sent me the passage money and says he'll try and get me a start on the railroad. As you know there wasn't much at home for me. Someone said I should sit for the new Civil Service Exam but I was never one for the books. Master Sweeney used to say I was a born clown and the first circus that came along I should join it. I hear they have brave big circuses in America.

I didn't come down to say goodbye. Things happened quickly in the finish and – I don't like goodbyes anyway. To tell you the truth, Bridie, I am terrified. I was terrified even travelling this far and I'm still only in Ireland. What will it be like over there in the middle of a big city full of all classes, creeds and colours and not one of them able to speak plain Monaghan English, as Barney McMahon used to say.

I hope you're not cross with me Bridie. You know how I feel about you. I hope you will write to me when I get my address in the States – The Main Street, Philadelphia – and that we will meet again some day. Say hello and goodbye to Sarah for me. I had a soft spot for her too but she made her own choice. I hope she's happy. Good luck to the lads at home too. And tell Canon McKenna there's no hope for me – I'm going to be a motion picture star. Think of me sometimes.

Yours affectionately,

Jimmy McC.

PS If you get a chance, drop in to my mother. She wasn't at all happy with my going.

J.

When she had finished reading the letter Bridie continued to stare at it until a tear rolled off her cheek and onto the page, smudging the word "affectionately." She gave an angry sigh, dabbed the smudge with the corner of her apron, wiped her eyes and carefully folded the letter before slipping it under the Sacred Heart statue on the mantelpiece. She sighed again and parroted his words, "You know how I feel about you." "Ye stupid ould clown," she thought. "It wouldn't have killed you to come out and say how you felt."

The kettle was purring over the fire. James put his head in the door.

"You smelt the tea-leaves," Bridie said, turning her face away.

"Any news?" James enquired. "I saw Owen Brady go up to the brae."

"Oh, plenty of news. Owen and Dev are going to start the Civil War all over again if Cullyboe is brought into the Six Counties." She plonked a mug of tea on the table. "And those cursed hens are laying out again. I'm off to look for eggs." She swept past her brother and was across the haggard before a bemused James had reached his mug of tea.

Sarah was never at ease in the presence of Winston Pollock. To sit beside his huge frame in the front of the Baby Ford was intimidating, but she had to be thankful for his offer of a drive to Newry. For some weeks now she had been concerned by the baby's movements. There seemed to be an irregularity, an inconsistency, which worried her. Gordon had suggested that she go back to Dr McKinney in Newry and had asked Winston if he would take her. Sarah extended her hands over her stomach and waited anxiously for the movements while she looked out the window.

"The countryside is looking very well," she said.

"'Tis that," Winston answered gruffly. "Who would want to give them wee hills away to another state?" he added sharply. "There's rumours going that the Boundary Commission is going to do just that. It's time James Craig told yon people that the loyal sons of Ulster'll not stand for that. And if he doesn't tell them we will, with our blood if necessary."

Sarah felt a shudder down her spine. She remained silent for a long time. As they entered Newry she felt sufficiently bold to speak again.

"It's great to be able to travel by car. It was very kind of you to bring me."

"'Tis only a Christian thing to do," he said.

Dr McKinney welcomed Sarah and listened patiently as she told him of her anxiety about the baby's movement.

"From what you say I don't think there's anything unusual amiss. I hear the same thing from most first-time mothers but just to make sure I'll have a wee listen."

He felt her swollen stomach and then listened through the stethoscope for what seemed an eternity to Sarah. At one stage he removed the stethoscope and was about to speak when he raised an index finger, smiled to Sarah, replaced the stethoscope and listened again.

"That's grand now, Mrs Johnston," he said finally as he folded the stethoscope and replaced it in its case.

"Everything is all right?"

"Your babies are grand."

"My – babies?"

"Yes. As far as I can establish you have two bonny babies on the way, Mrs Johnston. Congratulations!"

Sarah travelled home in a daze. Winston Pollock was content to hum the same tune over and over and made no attempt at conversation. She leaned her head against the window and closed her eyes. Twins. Two babies. How would she cope? What would they be? What would she call them?

"Twins! Well that's great news entirely. Are you feeling all right?" Gordon had come in tired after a long day working at the hay but this news had invigorated him.

"I'm fine. Dr McKinney says they're two healthy babies."

"Twins." Gordon savoured the word. "Never had the like in our family."

"Nor in ours."

"Well I hope they're two strappin' lads. I could do with the help."

The bell over the door announced Bridie McKevitt's arrival in Kernan's Bar, Grocery and Lounge. A flustered Podgy barely acknowledged her entrance. He wiped his brow with his shirt sleeve and faced his solitary customer.

"Is that it now, Aggie? Are you sure?" There was a note of exasperation in his voice. The slight figure in the long tweed coat mumbled a reply.

"All right. I'll go through the list one more time – "

Bridie smiled and moved into the cool darkness of the shop. This outer part served as grocery and bar. Beyond a glass-panelled door lay the lounge. She surveyed the grocery. Its untidiness betrayed Podgy's attempts to run the business on his own. When his mother was alive the shop had always been well stocked, the floor was swept a couple of times daily, glasses were properly washed. Since his mother's death a couple of years previously, the place had begun to fall apart –

"Cornflour? I thought you said corn plasters, woman." He searched a few shelves and finally held the cornflour packet up in triumph. "Now we're threshin', ses she, without the bags." He did a quick re-adjustment to the bill. "Seven and fourpence ha'penny, Aggie." She turned away from him before opening her purse and peering in. She extricated three half-crowns one by one and placed them on the counter. "And a penny ha'penny for yourself," Podgy said offering her the change. Again she turned away, secreted the three halfpennies in her purse, took up her basket and moved to go. She brushed past Bridie, not daring to look up but muttering as she passed.

"Tower of David, pray for us.

Ark of the Covenant pray for us."

As the doorbell signalled her exit, Podgy stretched his braces and gave a sigh. "If that woman isn't sent to Monaghan soon, I'll be in there before her!"

"What ails her?"

"Ach she was always odd as two left feet but in recent times she's gone to the bad altogether." He wiped his brow again. "This curse-of-God heat is killing me, and of course young Hanratty goes and injures himself playin' football and I'm left on me own while he goes to see the doctor. Will you have a glass of lemonade?"

"Were you playing football yourself?" she asked with a smile.

"Very funny. It's all this standing. I get a curse-of-God pain in me leg every so often. It's no joke runnin' this place on your own. You're at everyone's beck and call."

Bridie placed her basket on the counter.

"There's two dozen of the best eggs – and I want the best price for them. And I'll take a quarter stone of porridge oats."

"If that's the case you'll owe me money. The oul' eggs are very plentiful at the minute."

She thought of saying, "I suppose you want me to pay for the lemonade too," but kept her peace.

A gruff voice came from the lounge. "Begod, a man could die of thirst in here and never be noticed. Any chance of a bottle of Marcardle's this side of dinner-time, Podgy?"

Podgy muttered savagely under his breath and shuffled off to serve his customer. Bridie looked again at her surroundings. The floor was a mess of crushed cigarette butts. On an impulse she went through the brown-stained door to the kitchen, noticing the dented and scraped lower panel which Podgy kicked every time he went into his living-quarters. The kitchen was a shambles. The remains of breakfast scattered across the oil cloth on the table; a pile of dirty delph and pots on a smaller table by the window; two cats asleep in the hollow of a springless settee; wellingtons and shoes thrown under the table; a heavily laden clothes-horse in front of an ash-covered range. Bridie noticed a sweeping-brush beside the range. She negotiated her way across to the range, picked up the brush and hurried back to the bar. She gave a deep sigh as she began to sweep the floor.

CHAPTER EIGHT

BRIDIE SHOOK OFF THE SINGLE SHEET THAT COVERED HER. THE OPPRESSIVE heat that denied her sleep had lasted for a week now. She was not the only one in the house to have a sleepless night. Above her the floorboards creaked as James moved about in his little room off the loft. The familiar cough. Another cigarette lit. The clock in the kitchen gave four sluggish chimes. Already the first twitterings of the dawn chorus came from the hedge outside her window. She would get no more sleep now. Better to make a cup of tea. James would probably like one too.

She had scarcely reached the kitchen when a furious clattering of hooves down the lane told her something was wrong. Gordon reined in the horse with some difficulty and leaped from the trap.

Bridie was already at the gate.

"Sarah," he panted. "She's started – said to get you." He gulped a huge breath. "Can you come – now?"

Bridie nodded. She dashed up the stone steps to the loft and put her head in the door to tell James where she was going and why. Gordon spoke little on the way back to Sloe Hill except to say that he had already collected the midwife, Maggie Cooper.

"Where's Winston?" Bridie asked.

"Away for training with the 'B' Specials. Of all the times to be missing."

There was little practical help that Bridie could offer. Maggie Cooper had taken control from the moment she arrived and she bustled about efficiently, alternating a brusque order to Rose with a reassuring word to Sarah. Rose had to attend to her mother's calls so when Bridie entered the house there was a busyness about that seemed to exclude her. Sarah managed a wan smile of welcome. Her labour was long and difficult. It was made even more difficult by the oppressive humidity of a very warm summer morning. Bridie found a bowl and a hand-towel and began to wipe her sister's face and neck.

"Come on, now, love, you're doing just fine," Maggie urged. "For God and Ulster – push!" Bridie bit her lip at this remark and squeezed Sarah's hand.

The heat of the day had reached its peak when Sarah at last gave birth – a boy and a girl. When they had each been washed and wrapped in a cotton sheet, Maggie made two improvised cots from the top drawers of the oak chest that stood on the landing at the top of the stairs.

"That'll do them rightly till they get something better," she mused as she busied herself with her afterbirth duties.

"Winston Pollock was to collect a twin pram in Newry next week," Sarah explained.

"Sure you weren't due for another ten days," Rose interjected as she left to attend her mother.

"If you'd waited another few hours you'd have had two Orange babies," Maggie teased. "Tomorrow being the Glorious Twelfth and all. As it is – " her voice trailed off as she smoothed down the sheets, "as it is it's about time their daddy saw these two mites."

Gordon was initially embarrassed in the presence of four women but he could not take his eyes off the two infants now sleeping in the drawers. "The poor wee mites," he whispered. "We'll have to get Winston to collect that pram – "

"There's no need to rush," Maggie Cooper countered. "There's many had a worse start in life."

It wasn't until Maggie left with Gordon in the trap that Bridie had a chance to talk to her sister.

"How do you feel now?" she enquired.

"Exhausted! Whatever you do, don't start with two."

"Fat chance!"

"Any word from Jimmy?" Sarah asked. It seemed the obvious question. Bridie flashed her a look of disdain.

"No. Too busy conquering America. Have you done anything about the christening?" she added quickly.

Sarah was taken aback at her sister's tone.

"No. Would you arrange something with the Canon?"

"Surely. Have you chosen names for the pair yet?"

"We've talked about it – but nothing's decided."

"I hope there won't be any Winstons or anything like that." Bridie peered in at the tiny faces. "The wee girl has a look of Mother. High forehead and all. You'll surely call her Kath –" A loud choking sob startled Bridie. "What – what's wrong?"

Sarah shook her head and tried to compose herself. "Nothing," she whispered. "It's – nothing. Just a bit weepy – after all of this."

Bridie felt a sense of guilt. "You should be resting instead of listening to me prattle on. I'll go downstairs for a bit."

She went quietly to the kitchen, but finding no one there, she decided to stroll out to the orchard for a breath of air. The apple trees offered welcome shade from the burning sun. She leaned against a tree and caressed its bark.

"I thought you might like some tea in the open air." Rose's quiet voice woke her from her reverie. She put a tray down on the grass and began pouring the tea into delicate china cups.

"This is lovely," Bridie said, savouring a scone. "It's seldom enough I have a picnic these days. When we were children, we used to bring a picnic to Coomataggart for – " She stopped suddenly. Wrong subject. "The babies look grand and healthy, thank God."

Rose nodded in agreement.

"Would you like me to stay? You have your hands – "

"Not at all. We'll manage. Maggie Cooper will look in each day and Sarah will be fine after a rest. Anyway, you have your men to look after!"

"Oh indeed!" Bridie attempted a laugh. "Three more babies! I suppose I should get back or they'll die of starvation."

"You'll wait till Gordon comes back with the trap? He's gone over –"

"No, I'll walk. It's a grand day."

"It's very hot."

"I'll be fine. Thanks for the tea. 'Twas just what I needed. I'll be over in a day or two – if I may."

"Of course. Any time."

Bridie felt a sense of relief as she strode down Caraher's Brae. For Sarah's sake she would have stayed in Sloe Hill but she could never be at ease there. She said a little prayer that her sister would be all right. She was puzzled by the strained, almost frightened look on Sarah's face. Maybe giving birth did strange things –

"Halt! Password?" The dishevelled figure stood in the middle of the road, aiming his staff at her.

"Jemmy Patchy! You'd put the heart crossways in a body!" Then remembering Jemmy's affection for Sarah, she smiled up at his bearded face. "The password is twins, Jemmy, Sarah had twins today."

Jemmy jumped smartly to attention and saluted. "Very good, sir. I will relay that information to the Officer-in-Charge."

He clicked his heels and turned down the leafy lane to Coomataggart.

"Pack up your troubles in your old kit bag
And smile, smile, smile," he sang as he went.

Podgy Kernan paused, breathless, at the top of the loft steps. He stood holding the railing with both hands as he gulped air into his lungs.

"Are ye admiring the view, Podgy?" James called, gripping a row of brads between his teeth.

"View, my eye. Them steps'd kill a man. This curse-of-God foot is killin' me."

He hobbled into the loft and seated himself on a butter box that James used to store offcuts of leather. Podgy struggled to take off his left boot.

"Here, can ye do anything with that? Let it out a bit or something. Me curse-of-God ankle is bursting."

James noticed the bloated and bruised ankle as he bent to pick up the boot.

"Did ye not see the doctor about it?"

"Ah, bloody chancers the lot of them. Only in it for the money."

"Still – "

"I didn't notice Bridie around."

"She's over with Sarah." James offered Podgy a cigarette and both men lit up.

"Oh, aye. Busy times over there, what?"

"Surely."

"Sarah is a good woman. Yon Johnston got a good catch there."

James motioned to Podgy to move aside so that he could rummage in the box for a piece of leather. Podgy limped to the door and gazed into the hazy distance.

"Don't suppose you have your eye on anything yourself?" he said tentatively.

"Ah sure, who'd have either of us one-legged ould bachelors, Podgy?" James said as he slit the seam of the boot.

"Begod, there's a lot to be said for havin' a good woman about the house all the same," Podgy muttered, sucking furiously at the butt before flicking it into the haggard.

"Aye, the washing would get done at least," James said drily.

"Ah, there's more to it than that. A woman needs company too. And if a man can offer a house and a bit of money – "

James was not to be drawn. He waxed a length of thread slowly.

"Curse-of-God heat," Podgy muttered. "We need a good spill of rain."

There was a silence broken only by the light tapping of a punch through leather. Podgy grew fidgety, picking up and putting down tools on the workbench.

"I seen Hughie steppin' out with Josie McLarnon on Sunday."

"Mmm."

"Ah James. Ye know what I'm about. Would ye – is there any chance you'd put in a word for me – with Bridie?"

James stubbed the butt of his cigarette in an old tobacco tin.

"Would you not think of askin' her yourself, Podgy?"

"Ah sure I drop hints here and there, but she treats me as a joke. And then they say she's only bidin' her time for McCabe to come back from the States . . ."

He left the latter statement hanging as a question.

"Can't help you there, Podgy. Women are touchy on these things."

"Still you could put in a word?"

"I'll see what happens, Podgy, but I'm not promising anything."

"Well if you can, if you can." He swept a fly from his sweat-beaded brow. "Curse-of-God flies."

Brian and Hannah Johnston were christened after nine-o'clock Mass on Saturday in Cullyboe church. Hughie and Josie McLarnon acted as godparents for Brian. Owen Brady and Mary McCabe stood for Hannah.

"I would remind all here of the agreement entered into at the marriage of Sarah McKevitt and Gordon Johnston regarding the bringing up of their offspring in the one true faith." Canon McKenna dipped his head and peered over his glasses at the little group in the baptistry. "It is incumbent therefore on all of you to ensure that these children – who are now children of God and heirs to the kingdom of heaven – will never be deprived of that heritage. We will now have the churching ceremony for the mother." He motioned to Sarah to kneel at the altar rail while the others remained in the front pew. He fumbled through the liturgical prayer-book before solemnly intoning the churching prayers.

Outside, Winston Pollock drummed his fingers impatiently on the steering wheel of his car. He glanced anxiously at the few people that passed by the church at this early hour. There had been trouble at Cullyhanna the previous week – shots fired at a B Special patrol. He felt ill at ease in the Free State. What in God's name was keeping them? For all his vigilance he was taken completely by surprise by the figure who suddenly appeared behind him.

"Enemy behind our lines, sir!" The throaty voice froze Winston to his seat. He slid his hand under the seat where his Webley lay. As he did, the figure sprang forward to face him.

"Prisoner taken, sir!" Jemmy Patchy barked, never taking his eyes off the driver.

Winston Pollock seethed with rage and shame.

"Go away you bloody fool or I'll blow your brains out," he muttered. "Go on, go away!"

"Prisoner resisting arrest." Jemmy stood his ground.

Just then the christening group emerged from the church. Winston felt deeply embarrassed. "Does anyone know how to get rid of this idiot?" he asked nervously.

Sarah walked up to the ragged soldier. "It's all right, Jemmy. It's all right." She fumbled in her pocket and found a shilling and two pennies. "Here. Get yourself something to eat." She pressed the money into his calloused hand.

65

He saluted her and stood at ease. "Enemy repelled, sir. Lines intact!"

Winston Pollock swung the starting handle of his Baby Ford viciously. The engine shuddered into action immediately.

That night Bridie added a new prayer at the end of the rosary. "For Sarah and her babies – that they may grow in health of mind and body and in the fear and love of the one true God."

Pete scraped his chair noisily as he rose from his knees.

"Begod, them trimmin's is getting awful long, Bridie. I hope there'll be no more births, marriages or deaths for a while" – he looked deliberately at Hughie – "or we'll be on our knees all night."

Bridie was already dishing out the porridge even though James was still on his knees in private prayer.

"An extra minute on your knees won't harm you," she said. "You can take it off your football time." She searched anxiously in the drawer of the kitchen dresser.

"We're all business tonight and that's for sure," Pete said.

"Did anyone see the pen? I want to write a letter."

"Who's the lucky one?" Hughie asked between mouthfuls of porridge.

"Mary McCabe gave me Jimmy's address so I thought I might write him the news."

Pete, with his back to his sister, chanced a wink at Hughie. James remained kneeling.

CHAPTER NINE

116 Poplar Ave
Philadelphia
Pennsylvania
USA
Oct 4th 1926

Dear Bridie,

Thank you so much for your most welcome letter. I know you think the worst of me for not writing but it took me a long time to settle in here and to tell you the truth I'm still not settled and doubt if I ever will be. Give me Kerley's bog any day in preference to the busy streets of Philadelphia. It's great that all went well with Sarah. Please give her my best wishes.

Well here I am in the land of Uncle Sam working on the railroads. This is one big country, Bridie – too big for my liking. There's all colours and creeds here and that's for sure. You should see the traffic in this city. I can tell you Pete McMahon's dog wouldn't sleep in the middle of the road like he does in Cullyboe.

The work is hard on the back but it's work and it's money. Give me five years and I'll be President of the railroad! You never know who you'd meet out here. I went for a drink in a bar and wasn't the barman from Dundalk. Knows Podgy Kernan well. You should see the bars they have here – tell Podgy he'll have to come out and see for himself! Me and another fellow from Galway went to "the movies" to see a fellow called Buster Keaton. It beat all. We were in fits laughing.

I suppose you think I'm having a brave wild time over here. To tell you the truth never a day goes by when I don't think of Cullyboe – and you. Little things – like the schoolhouse dance or the smell of the flax or sitting on Pete McMahon's window after Mass. I look down that never-ending railway line and all I see is Cullyboe with its wee hills and fields. It's a terrible thing to be so far away from your own folk. Terrible. I must go now. Remember me to all.

Think of me sometimes,

Jimmy.

Bridie slipped the letter into the pocket of her apron. She quickly turned away on hearing James limp past the scullery window.

"I thought I'd make a cup of tea," he said on entering. "It's brave cold up in the loft – "

"You take a turn with the churning and I'll get it," Bridie said, keeping her back to him. Her voice was little more than a whisper.

"Not bad news I hope. I saw Owen Brady go past – "

"No. No." She adjusted the crook on which the kettle hung over the fire. "It was a letter from Jimmy – "

"There'll be a blue moon tonight, surely!" James attempted a chuckle as he began turning the churn handle.

"He's doing fine. Asking for you all." She took two cups from the dresser. The churn tumbled slowly, its contents sloshing noisily with each turn.

"Podgy Kernan was asking if you had any eggs for sale."

"I haven't."

"To tell the truth he was more interested in seeing you than seeing the eggs."

"Oh – does he need the kitchen cleaned up again?"

James stopped, removed the lid of the churn and peered in. The first globules of butter bobbed about on the top of the cream. "Podgy's a dacent enough ould sort, Bridie," James said into the churn.

"The tea's made," his sister replied curtly. "I'll finish that out."

Rose Johnston stormed into the kitchen. "Are you deaf or what?"

Sarah clutched her tweed coat over her nightdress and stared dreamily at the upper shelves of the dresser.

"I said are you deaf – " Rose gritted her teeth with impatience.

"Is there no biscuits?" Sarah asked.

"Biscuits? Biscuits! Your babies are bawling with hunger and you're worried about biscuits for yourself!"

"They can't be hungry. I only fed them a wee while ago – "

"You fed them hours ago. Hours ago. One of them's been crying for twenty minutes. Now the other's awake. I'm trying to keep Mother quiet. She's all upset thinking it's her own babies. Will you for God's sake do something with them?"

Sarah turned and faced Rose with a smile. "I'm tired, Rose," she said softly. "Are you sure there's no biscuits?"

For a moment Rose stared at her sister-in-law, incredulous. Then with tears welling in her eyes she swept past Sarah into the pantry in search of the muslin-covered jug of milk.

Bridie took the withered potato stalks in her left hand and shook them vigorously. Her right hand probed in the soil for the pink-hued potatoes. Some twenty yards ahead of her Hughie and Pete each dug out a drill of potatoes. Hughie paused and cast a worried look at the lowering sky behind them.

"We're in for another belt of rain. We'll not get this lot finished the day."

Pete rested on his fork. "If we had our full team out we'd beat the rain and no mistake," he said. "D'ye mind last year we had Sarah – a great wee lassie at the spuds?"

"Aye!" Hughie resumed digging. "And we had Jimmy McCabe too – for all the good he was. Acting the clown with Bridie most of the time."

Instinctively, both men looked down the sloping field at their sister. She gave them a sullen look before turning with her full bucket back to the sack that squatted further down the slope.

"For God's sake," Peter whispered. "Be careful when you mention Jimmy. She's very touchy about him at the minute."

"She's very touchy about everything, if you ask me. She nearly ate me without salt yesterday when I asked her if she'd ask Podgy Kernan if he'd let us have young Hanratty for a day at the spuds."

"Well that's what's known as puttin' your two big spawgs into it!"

Hughie looked at his brother with genuine puzzlement.

"How so?"

"How so! God, Hughie – did you come down in the last shower? Isn't Podgy pestering her ever since Jimmy went to the States! And anyway Hanratty would be as useless out here as he is on the football field. I could have brained him last Sunday – spent the whole match slickin' back the hair and eyein' the girls on the sideline. Hardly got a kick at the ball!"

Hughie chuckled as he sliced the fork into the drill and upturned another clump of potatoes. He prodded them loose from the soil. "Thank God for a good crop of Pinks anyway," he sighed.

Bridie returned with an empty bucket and bent to her task once again. Pete was breaking down another clump when he noticed a figure cycling along Flaxpool Lane towards the McKevitt house.

"That's hardly Hanratty now," he said, straining to see in the fading light. "Although it would be just like him to come now when – "

Hughie peered into the gathering gloom. "Begod and it's not Hanratty. It's a woman. I'd swear it's Rose Johnston."

James was wrestling with the stitching of a horse-collar when the timid knock came on the door.

"Come in! Come in! We're open for business." He was taken aback to see Rose Johnston standing in the doorway.

"I'm sorry to interrupt the work. There was no one down below – "

"They're away at the spuds. Come in out of the cold, Rose. There's a bite in that wind."

She closed the door behind her.

"Well, we don't see you too often in these parts!"

Rose smiled nervously at him. "It's about – "

"Is there something wrong – "

" – Sarah."

He motioned to her to sit on the tall stool by the workbench as he fumbled for his cigarettes.

"Is she sick?"

"No – she's not ailing as such. I wish I knew what's wrong with her."

James lit a cigarette and drew deeply on it. "How do you mean?"

"It's just – just as if she doesn't care any more. She doesn't seem interested in the babies. They cry a lot. She has to be reminded to feed them."

"Did you say anything to her?"

"Of course. But it makes no difference. She's – in another world, almost."

"What does Gordon say?"

"Gordon says – " she began to sob quietly. "Gordon is Gordon. He doesn't say much. He suffers on. And then there's Mother. She's getting worse." Rose stifled a sob.

James shifted uncomfortably on the upturned box and stubbed the cigarette under his heel. As he reached again for the cigarette box, Bridie entered the loft. She nodded nervously at Rose. James told her what Rose had said.

"What about the doctor? Has the doctor seen her?" Bridie scraped the clay from her nails.

"No. She – she's not really ailing. And anyway, she wouldn't go. She doesn't see herself as – as needing a doctor."

Bridie sighed. "I'll go and see her – tomorrow. There's hungry men to be fed here. Will you take a drop of tea?"

"No. No thank you. I need to get back. Gordon is – you'll be welcome tomorrow."

She opened the door into the November evening and hurried quickly down the loft stairs.

Canon McKenna had just poured himself a large neat whiskey when a loud knock on the back door of the presbytery interrupted his reading.

He muttered to himself about callers at this time of night as he made his way through the kitchen to the back door. When he opened the door, an icy blast of wind attacked the flame of the oil lamp which he carried in his left hand. A tiny bird-like figure, wrapped in a long black cloak, stood on the doorstep.

"Aggie Trainor, what ails you at this time of night, woman? Is there someone ill?"

"No, Canon. I need – " her frail voice was almost drowned by the howling gale.

"Come in, woman. Come in or we'll both get our death." There was an impatient edge to his voice. He motioned towards a chair. "Now sit there and tell me what ails you." He thought of returning to the parlour for his whiskey but decided that this would be a brief call.

"I need you to give me a penance, Canon."

"Penance? Weren't you at confession on Saturday night as usual?" His annoyance was growing but he made an attempt at humour. "Don't tell me you've murdered someone this black night!"

"The world is full of sin, Canon. Full of the most awful sin. And the voices are telling me I must do penance." She broke into a monotone chant.

"Tower of ivory, pray for us.

Ark of the covenant, pray for us.

House of gold –"

"The voices, Aggie?"

"The voices of the souls in purgatory, Canon. Like you said at Mass. The voices of the holy souls are crying out. I must do penance for them – "

"And you came here at this hour of night – " He checked himself. Anger would not remove this woman. "Stay there," he commanded. He went back to the parlour, took a long gulp of whiskey and searched on his desk for a small bundle of leaflets. He selected one. "Now, Aggie, there's a special prayer to Our Lady of Fatima. I want you to say it three times a day for the conversion of sinners."

"Our Lady of Fatima? Will that be enough, Canon? There's so much black sin in the world – "

"There is indeed, Aggie. And only our constant prayer to God and his holy mother will overcome it. Off you go now. Home to your bed and say that prayer." He opened the door as he spoke.

"I will, Canon. I will." She clutched the leaflet to her breast as she faced into the biting wind.

"Our Lady of Fatima, pray for us

Queen of sinners, pray for us."

Canon McKenna hurried back to the warmth of his parlour, drained his glass of whiskey and quickly refilled it. He paused for a moment, then sat at the mahogany desk and withdrew a leather-bound pad from a drawer. He savoured another large swallow of whiskey before reaching for his pen. He would write to Pat Trainor straight away.

CHAPTER TEN

BRIDIE GRIPPED THE COLLAR OF HER COAT ABOUT HER AND, WITH SOME difficulty, wheeled her bike one-handed up Caraher's Brae. The northerly wind swept down the brae, needling her face and causing her eyes to water. She bent her head into the gale.

Her mind was in a confused state, Rose's words troubled her. Why would her sister "not care about her babies"? How could she be like that and not be ailing? Bridie's confusion was compounded by her meeting with Podgy Kernan. She had called in to the shop on her way to Sloe Hill to buy a couple of chocolate bars for Sarah. It would save her going into Cullyboe.

Podgy emerged from a shed behind the house wiping his hands on an already much-stained collarless shirt.

"Well you're the early bird! Bottlin' that stout is a curse-of-God nuisance. You're welcome! You're welcome!" He spoke quickly, betraying some nervousness.

"It's not a social call, Podgy. I just want a couple of chocolate bars."

"Aye. Come in this way." She followed him through the kitchen. "I didn't open the shop yet for fear of Aggie Trainor."

"For fear of Aggie?"

"Och, the woman's gone totally demented. Sell her a bar of soap and you get the rosary and the secret of Fatima in return."

Bridie smiled at the thought.

"Sure she's a harmless soul for all that."

"Harmless? Shoo!" He aimed a kick at the cat that lay stretched in

73

his path. "Between herself and that eejit Jemmy Patchy marchin' in here and orderin' us all to take cover 'cos the Bulgars are comin' down Caraher's Brae – I'm totally tormented!" He struggled with the lock of the door into the shop. "Totally tormented. Is it the curse-of-God weather or what that drives them to that state?"

"I don't know, Podgy," Bridie laughed.

"Well begod I don't know when I saw you laughin' last, Bridie," Podgy chuckled. "It suits you."

"I'll take two bars of chocolate, Podgy." Bridie adopted a more serious tone. He turned his back to her and scratched his head as he searched the shelves for chocolates.

"Aye. There must be hope for me if I can give you a laugh. What would you say, Bridie?"

There was no reply.

"Where did Hanratty put the curse-of-God chocolate? Probably ate it all himself."

"It's beside the tea."

"So it is. So it is. Begod, if I had you here, Bridie, the place would be ship-shape in no time at all."

She fumbled in her purse for a sixpenny piece and offered it to him. To her surprise he took her hand in his.

"I mean it, Bridie." His rough hand tightened on hers. "You're a good tidy woman. And I could give you a good livin'." His face puffed red with embarrassment.

"I have to go, Podgy." She managed to free her hand from his sweaty grasp.

"You'll waste your life if you wait for McCabe. He'll not come back." There was a harder edge to his voice.

"Nobody's waiting for anyone," Bridie snapped as she made for the kitchen door.

Podgy was taken aback.

"I'm only sayin' you'd be welcome in this house," he said sheepishly.

"This is not the time to be saying anything, Podgy." She swept through the back door. Podgy tripped over the cat.

"Shoo! Ye curse-of-God nuisance!"

Rose opened the door to Bridie.

"I saw you from Mother's window," she explained. "You must be frozen. Come down to the kitchen. Sarah's there."

"It's Bridie, Sarah."

Bridie nodded with a smile to her sister. Sarah smiled and looked away through the window.

"I'll take your coat," Rose said. "Put on the kettle, Sarah. I'm sure you and Bridie have a lot to talk about."

"Well, how are the weans?" Bridie asked tentatively as she warmed her hands before the fire.

"Oh the weans, the weans," Sarah replied dreamily. "The weans are asleep."

"Probably get more sleep than their mother," Bridie joked.

"Wouldn't be hard." Bridie noticed that her sister still wore her nightdress under her tweed coat.

"Would you like to have a wee rest now while I'm here? I could – "

"What would I do that for? Sure, I'm their mother. Gordon has his farm. Rose has her mother. And I've got the weans."

Bridie filled the kettle and hung it over the fire.

"Well, we'll have a nice cup of tea and look – I've brought you a couple of bars of chocolate from Podgy's!"

"Is Podgy well?"

"Well as he'll ever be. Complaining all the time."

Rose came to the door.

"The twins are crying, Sarah."

"They're always crying. They'll wait a while – "

"I'll go to them," Bridie interjected. "I'm dying to see them."

Sarah unwrapped her chocolate bar.

Rose led Bridie up the stairs and indicated the room where the twins lay.

Bridie was taken aback at the state of the room. Although it was small and sparsely furnished, it was an unkempt mess of baby clothes – both clean and soiled – basins and a bucket of fetid water. Worse was the smell that confronted Bridie. Both babies were in distress and clearly had not been changed for some time. Bridie thought of turning back in anger to her sister, but instead closed the door behind her, wiped the tears from her eyes, rolled up her sleeves and set to work.

An hour later Bridie carried two buckets across the haggard to a paddock behind the cowshed. She made her way to a deep ditch into which she emptied the buckets. She hurried back across the paddock. A voice called to her from the cowshed. It was Gordon. He summoned her to the shelter of the shed.

"Well," he asked, propping himself on the fork he had been using to clean out the stalls, "what do you think?"

Bridie shook her head. "She's not well. The state of that room. Of the babies. Sarah was never untidy like that."

"Don't I know it?" He poked at a wisp of straw with the fork. "What am I to do?"

"She should see the doctor. Maybe she needs a tonic of some kind. She can't go on like this."

"I'll ask Winston if he'll take her to Newry on Wednesday."

"I'll come over and mind the twins – if you want."

He nodded.

"I think maybe you should go with her, Gordon . . . "

"We'll see."

On Wednesday afternoon Bridie waved goodbye to Sarah as she left for Newry in Winston Pollock's car. She choked back a sob as she watched the tiny bird-like figure wave back to her. It seemed to Bridie that her sister had shrunk in size in the previous months. The green tweed coat that they both had chosen for Sarah only a year ago now hung loosely and without shape on her. Gordon waved nervously from the front seat.

"Have you laid the table yet?" the tiny voice called from the bed.

"I'm doing it now, Mother," Rose called back from the hall below.

"Make sure it's the blue china you put out. The Reverend Nelson doesn't come very often and Mrs Nelson is so particular. Is the fire lighting in the parlour?"

"Yes, Mother."

"You're a good wee girl, Rose. Is Gordon home from school?"

"Not yet, Mother."

"Well send him straight up to me when he comes. I want to make sure he changes into his Sunday suit. I hope Harold gets the milking done in time. I don't want him coming to tea smelling of cows –"

Bridie fed the twins and wrapped them up warmly. She decided to take them for a walk. A watery sun hung low in the sky as she wheeled the pram out the gravelled drive and down the Cullyboe road.

"This'll put a bit of colour in your cheeks," she cooed at the two wondering faces. "And it will keep me out of the way for a while. And then Mammy will come home in much better form . . ." She responded playfully to the gurgling of the infants. An old man on a bicycle saluted her as she passed. She was just about to turn back when the dishevelled figure of Jemmy Patchy appeared from a gateway. He stood to attention.

"All quiet, sir. Nothing to report."

"That's good, Jemmy," Bridie smiled. "I have two new recruits to report. Sarah's babies!"

Jemmy peered cautiously into the pram. Brian screwed up his face at the sight of a bearded stranger. Jemmy saluted quickly and clicked his heels.

"I will advise the men of the reinforcements, sir."

Bridie turned for Sloe Hill. "You do that, Jemmy. You do that."

He marched off in the opposite direction, staff on shoulder, chanting as he went.

"Left, left, left right left . . . "

On her return Bridie busied herself with a surprise for Sarah. She had brought the ingredients for a Christmas cake and had only just got the mixture into the tin when she heard Winston's car arrive. She feigned surprise on seeing Gordon and Sarah enter.

"Och, another five minutes and I'd have had your surprise in the oven – but now that you're here let ye each give a stir for luck – and make a wish."

Gordon was genuinely pleased. "Isn't that lovely, Sarah? I love sweet cake," he exclaimed with boyish innocence. Sarah smiled at him as she slowly stirred the mixture.

"Brian and Hannah and me had a lovely walk," Bridie announced. "They'll sleep well after all that fresh air!"

"You're very good, Bridie," Gordon replied. "We'd be lost without you, wouldn't we S- ?" Sarah had slipped out of the kitchen while Gordon took his turn to stir.

"Well," Bridie whispered, exploiting the opportunity, "what did he say?"

"Och, not a lot. He says it's just something that happens. That she'll come right in time. Gave her a tonic. Told her to get out in the air more. Go visiting." He stopped stirring but still gripped the spoon tightly. Bridie sensed the lack of conviction in his voice.

"Maybe if I came over – more often. Once a week?" she suggested.

"Maybe." He shook himself out of his reverie. "The milking will be late this evening."

Bridie was surprised to see light coming from the loft on her return home. James did not normally work this late. She decided to investigate. James was crouched in familiar pose, cigarette dangling from his lips as he stitched the upper of a boot. She noticed how he worked close to the oil lamp.

"Och, James, you haven't the light for work like that," she chided him. "What has you up here at this time of day?"

He ignored her question.

"How did Sarah get on?" he asked.

"The doctor says she'll come out of it. She needs to get out a bit more."

"How is she to do that?"

"I offered to go over once a week. Now what has you – "

"We have a visitor. Josie McLarnon."

There was an uneasy silence.

"Did ye all run at the sight of her?"

"Well, Pete's gone to a meeting of the football club and I – had a bit of work to do – "

"In the cold and the bad light? Come on down to the kitchen, for God's sake. She'll not eat you!"

Bridie rattled her bike intentionally over the cobbles of the haggard and took a deep breath in the porch before entering the kitchen with a broad smile.

"Well, hello Josie. You're welcome!"

"Thanks, Bridie. I was – just about to leave." Josie's nervousness caused her to knock over the fiddle that rested against her chair.

"What would you be doing that for? It's early yet and I see you have the fiddle with you. We'll have a tune out of you before the night is out. Did they make you a cup of tea itself?" She looked directly at Hughie who grinned sheepishly at Josie as he scratched the back of his head.

"Lord save us from lazy men," Bridie sighed. "Leave them on their own for a few hours and they can do nothing for themselves. Put on that kettle this instant, Hughie McKevitt." Hughie jumped into action.

"How is Sarah?" Josie asked. "I heard she wasn't –"

"She's rightly," Bridie interrupted. "She just needed a tonic."

James came in, rubbing his hands.

"There's a brave cold wind out there," he said as he drew up a chair to the fire.

They talked cautiously at first of happenings local and beyond.

"I hear they're not going to change the Border," James said. He lit a cigarette from an ember he lifted from the fire with a tongs.

"I'm glad of that," Josie said. "I don't want to be in with them – " She checked herself, realising she might offend the McKevitts.

"Who would?" Bridie said. "Who would?"

"Still and all," Hughie said, "I don't think them Free Staters know what they're at, at all. De Valera would make a better fist of things – "

"You should hear Owen Brady on about Cosgrave," Bridie laughed. "Owen's all on for starting another war!"

"Is there any news of Jimmy McCabe?" Josie asked. It seemed an obvious question at the mention of the postman but for the second time she realised she had said the wrong thing.

"Och, building railroads halfway across the States," Bridie said, reaching for the tea-pot.

"You'd miss him about the place," Hughie said.

"Aye," Josie agreed. "He was great at the dances."

"The tea is made. Let ye sit over to the table," Bridie commanded. "I suppose Pete will be talking football till midnight," she added.

They sat at the table and ate in silence, until Josie spoke in praise of Bridie's baking.

"That's a gorgeous cake, Bridie. I wish I could bake like that!"

"But you can make sweet music on the fiddle," Hughie said gallantly.

"Aye. Why don't you give us a tune, Josie?" James asked. "God knows when there was music in this house."

Josie obliged with a selection of reels followed by a slow air.

Bridie gazed into the blazing fire. The music took her back to a summer evening when times were happier . . . when Sarah smiled in Gordon's arms . . . when Jimmy McCabe sleeked his quiff and swaggered in front of the women during the ladies' choice, pleading with them not to rush all together . . . So much had changed and so much would change in the McKevitt household.

CHAPTER ELEVEN

As she sluiced the babies' bottles, Sarah allowed the water to trickle idly through her hands. She stood the bottles on the windowsill. Clean bottles. Bright clean bottles. She dried her hands on the roller towel on the back of the door. From the bottom of the dresser she took an old biscuit tin. She took Bridie's cake from the tin and wrapped it in a newspaper. She buttoned her coat over her nightdress right up to her collar and tucked the cake under her arm. As she lifted the latch on the back door the chimes of eleven o'clock echoed through the house.

The ice cracked under her feet as she made her way up the driveway. The cold numbed her feet through her light shoes. She walked a short distance down the roadway before turning down an overgrown boreen. She was oblivious of the tangled briars that snapped at her hair and snatched at her coat. She hurried along until the boreen suddenly opened out into Kerley's bog, a small area of bogland that provided turf for the homesteads in the surrounding townlands. Sarah followed a rough track that skirted the bog, startling the occasional snipe that shot out of the rushes on either side of the track. The bright morning sun silvered the dark pools and channels that scarred the bog. Bright clear water. The track petered out. Sarah continued across the fields. By now the hoary grass had soaked her shoes completely. She emerged onto another laneway and followed it until she came to a familiar building – an old ruined house that was used to store hay. She called out.

"Jemmy? Jemmy Patchy? Are you there, Jemmy?"

She waited anxiously until the familiar bedraggled figure appeared from the back of the building. He saluted.

"Present, sir."

"I brought you a present, Jemmy. A Christmas cake!"

She offered him the parcel wrapped in newspaper. He accepted it, somewhat bemused.

"It's because you are my friend, Jemmy. Happy Christmas, Jemmy Patchy!" She gave a little laugh. "Goodbye, Jemmy!" She skipped away from him as he saluted.

"I'll present this to the quartermaster and see that the men get their share, sir. Happy – "

She was already out of sight.

The Fadden river swirled through rushy fields, swollen by winter rains. The Fadden constituted the Border for a few miles of its meandering course. Sarah doubled back across the fields and came at last to where the water seeped over the low-lying grass. Bright and clear. The water was so clear, she could see the grass waving beneath. Water was cleansing. I baptise thee in the name of the father . . . Water was healing. Water was bright and clear. Sarah walked onwards. The weight of her sodden tweed coat dragged at her, as she trailed her fingers through the bright clear water. Water was . . .

Larry Markey drove home the peg of a rabbit snare with the heel of his boot and rearranged the snare carefully across the rabbit run. "Come on out now, yis little gets," he muttered. "Come on out for a run in the sun!" He picked up the remaining three snares and blew fiercely through his cupped hands. "Though I wouldn't blame yis for not coming out in this weather. Man dear, but it's cold!" He trudged onward through the glistening grass and sedge to the bottom of the field. The flooded river spread into the field beyond. "No point goin' in there," he thought. "Might as well peg these few – " His attention was drawn to a bloated green form that moved slowly to and fro, trapped between clumps of rushes. At first glance it seemed like an animal carcass but a horrible fear gnawed at his gut when he realised what he was watching.

"O Holy Mother of Christ!" he shouted, dropping the snares and clambering over the hedge. He splashed through the gently lapping water to the rushes and grabbed frantically at the green tweed coat.

The lifeless body spun around and Larry Markey found himself staring at the dilated eyes and gaping mouth of Sarah McKevitt.

"O sweet Christ!" he cried. "O sweet Christ!" He dragged the body through the water and back to the hedge. His own body shook with terror as he repeatedly blessed himself. "Sweet Jesus have mercy on her soul," he stammered under his breath before he burst his way through the hedge and made his way blindly up the field to the gateway where he had left his bicycle.

The biting sleet stung Pete McKevitt's face as he drove the trap home. He cried bitterly, loudly into the northerly wind. Behind him, James sat to one side, his sister's body stretched at his feet. He held her right hand in his, caressing it with his left hand as if he were reassuring her that all would be well and they would be home soon. Occasionally his whole body shuddered and he gave a deep sigh.

Dr O'Reilly had examined Sarah's body and confirmed death from drowning. Canon McKenna was away for the day. Hughie had gone with Dr O'Reilly to break the news to Gordon. James and Pete decided to take their sister the short distance to her home. As the trap rounded the corner at Flaxpool Lane, it nearly ran down the bent bedraggled figure that trudged up the middle of the narrow road.

"Get out of the way, ye bleddy oul' eejit!" Pete roared at him as they swept by. Jemmy Patchy stumbled to one side and, as he picked himself up, he caught a glimpse through the wildly swinging rear door of the trap of a body wrapped in a familiar green coat. He sat back on the roadside bank, put his head in his hands and began to cry.

Bridie was waiting for them at the gate, clutching her coat about her in the bitter cold. She uttered a loud sob on seeing her sister but quickly composed herself and issued orders to her brothers.

"Take her to our – to my room. Mary Ann Markey is here and she'll help me lay Sarah out." Mary Ann was Larry Markey's mother and was one of the first in the parish to know of Sarah's death. She was a hard, practical woman who would be helpful at a time like this. Pete shook his numb hands and snuffled and sobbed as he went to help James unload the body.

"Hurry now," Bridie said firmly. "I don't want Gordon – or anyone – to see her like this."

Her wish was achieved, just in time. Gordon arrived with Hughie in Winston Pollock's car just as Bridie smoothed out the starched sheets on Sarah's bed. Mary Ann stood back from the bed.

"She's lookin' right beautiful in that dress," she said. Bridie had decided to lay Sarah out in a long-sleeved dress with a pretty floral pattern. It had been a favourite of Sarah's which she had never brought to Sloe Hill, mainly because it did not fit her the previous summer when she was pregnant.

Gordon burst into the bedroom, his face white with shock, his eyes wild with fear at what he might see. He gave an angry choking cough and fell on his knees at the side of the bed. Bridie motioned to the others to withdraw.

When he eventually came out to the kitchen, Bridie gave him a cup of tea. The cup rattled off the saucer as he fought to control the shake in his hand.

"We brought her here because it was near," Bridie said quietly. "I hope you don't mind – "

Gordon shook his head.

"It's best," he whispered. "I suppose she'll be – buried from Cullyboe?"

"I suppose – "

"What happened her, Bridie? Why did she do such a thing?" His plaintive tear-filled eyes looked up at Bridie.

"Only God and herself knows that."

"The weans. What'll I do with the weans?" he sobbed.

"We'll find a way," Bridie said, filling his cup again.

Canon McKenna arrived in the afternoon and having expressed his sympathy to all present, he prayed for some time over the body.

"I would not be in favour of her laying-out apparel," he said, taking a cup of tea from Bridie. "It seems more fitting for a wedding – and given the circumstances of her death – "

"She was happy in that dress, Canon," Bridie said coldly. "She'll be buried in it."

The rebuff hurt him.

"Of course she'll have to be buried in the cillín. She can't be buried in consecrated ground."

There was a long silence before he added, "I have prayed that God in his mercy will forgive her."

At that moment Podgy Kernan stumbled in the door carrying a large wooden box which he deposited on the kitchen table with much rattling. He wiped the sweat from his brow, wiped his hand on his coat, clumsily shook Bridie's hand and gulped a big breath.

"I only heard an hour ago. What in the name of Christ – begging your pardon, Canon," he panted, realising in whose company he was. He gestured towards the box.

"I brought some stout and whiskey. You'll be needin' them – "

"I will take my leave now, Bridget," the Canon interrupted, peering into Podgy's box as he moved towards the door. "I hope this won't lead to over-indulgence," he said drily, looking straight at Podgy.

"Oul' hypocrite," Podgy muttered as Bridie closed the door. "Gone home now to swally a bottle of Black Bush. Wouldn't even buy the curse-of-God stuff locally. Has to go to Newry for it." He moved to the door himself. "I have a few loaves and a shank of bacon out in the trap."

Through the evening a stream of callers came to the house as word of Sarah's death went around the parish. Bridie was glad of Podgy's foresight in providing food and drink. Josie McLarnon proved a great help in making tea and sandwiches. At regular intervals one of the women would lead the saying of the rosary in the bedroom. The women knelt around the bed while the men crowded around the doorway. After each rosary the neighbours broke up into little knots and talked in hushed voices. Words of sympathy were exchanged awkwardly and embarrassedly with the McKevitt family who coped with their grief in different ways. Bridie worked ceaselessly, preparing food, washing and drying dishes. Hughie joined in every rosary and found comfort in the presence of Josie McLarnon. James was never without a cigarette and made a point of greeting every visitor. And, all through the long night, the inconsolable Pete sat in the bellows chair by the fire, wringing his hands, staring into the dancing flames and regularly breaking from a quiet whimpering into long angry sobs.

Sarah McKevitt was buried in the *cillín* – a corner of a large field across the road from Cullyboe cemetery. The graves of unbaptised children were marked only by the occasional rock that stood askew in the grass. A flurry of snow descended on the small gathering of people who had followed Sarah's coffin – borne on the shoulders of

her husband and her brothers – from the nearby church. James led
the recitation of the rosary as Hughie and Pete filled in the grave. The
snow became heavier and, when the McKevitt family finally withdrew
and made their way back to the trap in the churchyard, a fine mantle
of snow covered the new grave.

When all had departed, a cloaked figure emerged through the
swirling snow, marched briskly to the *cillín* gate and saluted in a
slow deliberate movement. He called out the order, "Reverse arms!"
and proceeded to bring his wooden staff slowly into the reverse
position. For two hours Jemmy Patchy performed a slow march, arms
reversed, up and down the snow-covered roadway outside Cullyboe
cillín.

116 Poplar Ave
Philadelphia
USA

My dear Bridie,

I am posting this good and early in the hopes that you'll have it in
good time for Christmas. Don't give Owen Brady any Christmas drink
until he delivers this letter!

Well, America is slowly getting used to Jimmy McCabe President
Coolidge invited me up for a bit of Christmas dinner, but I told him
Aunt Maisie got her spake in first . . . To tell you the truth, I'm
dreading Christmas. I've never been away from home for Christmas
before and while I won't want for anything in Maisie's, it'll not be the
same. I'll miss the goose and the feed of good Monaghan spuds (you
should see what passes for spuds over here). I'll miss the games of
cards and the house dance – and the chance of my arm around your
waist! And I'll miss the free drink in Podgy Kernan's – you tell Podgy
he's to keep that drink safe for me. He's not getting off that light.

The cold here is a total dread. I remember as a gossoon getting
the hands frozen off me washing the spuds for the pigs' pot but this
place beats all. We have about two foot of snow at the moment –
great weather for building railroads! The work is hard but I'm doing
well – saving the few dollars. Not easy when these American lassies
are always throwing themselves at me! Don't worry, I haven't fallen –
yet. Canon McKenna would go out of his mind here – a Godless
country, he'd call it. Myself and this McMahon fellow from Longford

went into the city the other night and this big lassie came up to us and said, "Hey honey – two dollars and I'll give you a good time!"

"Oh begod," says McMahon, "you'd have to pay me more than that!"

Enough of that. How are things with you? I hope well. Are you still having trouble putting manners on them big brothers of yours? Tell Pete I'm sorry the football team lost the Feis Cup Final. They obviously missed me! D'ye mind the time the *Democrat* wrote about me – "McCabe was industrious as ever and floated over two delightful minors . . . " I still have the cutting somewhere.

And how is Sarah? And the weans? I'll bet they're keeping her busy but I'm sure she's enjoying every minute of it. They must be fine babbies by now. Isn't she well away this Christmas? As for me, I'll be glad when it's over. I have to stop, Bridie. I pray you are well. Fire an ould snowball at Podgy for me.

God bless,
Jimmy McC.

CHAPTER TWELVE

"COME IN! COME IN!" PODGY KERNAN CALLED FROM THE KITCHEN. "I seen you coming."

Bridie entered from the shop.

"I'm in the middle of a job here and I'll have to finish it out." The job was cutting his toenails. Podgy rested one foot in an enamel basin of soapy water while he propped the other foot on the rung of a chair. He leaned forward and, with some difficulty, reached the toes of the dry foot with a scissors.

"I can wait in the shop," Bridie said, barely concealing her distaste.

"Not at all, not at all, I'm nearly finished. That curse-of-God big toe is the divil to manage." A splinter of nail pinged off the basin. "There!" he sighed. "That's the job!" Red-faced from his exertions, he smiled up at Bridie. "I'm sure you've seen a man's foot before," he said.

"Not one like that in the basin! You should get that seen to."

Podgy withdrew the badly swollen foot and dried it with a threadbare towel. "Hah! And pay good money to some ould quack that knows nothing! No bleddy fear. Salty water is the only man for that. Maybe when the weather is warmer, I'll take a trip down to Blackrock on a Sunday for a paddle in the sea. And maybe you might come yourself?" Bridie did not acknowledge his offer.

Podgy slipped the bare foot gingerly into his boot. "Now, ses he, we're in business. Will ye have a cup of tea?"

"No thanks," Bridie said. "I just need a few things."

"Suit yourself!" He shuffled after her back into the shop.

"How is things?" he asked tentatively.

"Good as they'll ever be." She looked at her list. "Have you any dry mustard?"

Podgy scratched his head as he scanned the shelves.

"God I don't know – "

"It doesn't matter," Bridie said.

"No, I mean I don't know how yis can bear that cross. Such a grand wee *girse* . . . Begod I have the mustard and all," he added with some surprise.

"And a half pound of tea." Bridie paused, then continued with a sigh. "We can only do our best. It's three months now and I still hear Pete cry himself to sleep like a baby."

"And yourself? What about yourself?" There was a genuine concern in Podgy's voice.

Bridie acknowledged it with a weak smile.

"I just pray that God will – I just pray."

"And what about Sloe Hill?"

Bridie shook her head.

"Gordon's a broken man. He just doesn't talk. I fear he may be looking for comfort in the bottle. I feel sorry for Rose. She has the weans and her mother."

"I thought you were helping out."

"They tolerate me one day a week. I need two boxes of matches and forty Players. James is smokin' a terror these days. And it Lent and all."

Hughie McKevitt talked gently to Major as he guided the plough on the turn at the headland on the top of the Hill Field. The horse seemed to nod in reply as Hughie pulled hard on the left rein and prepared for the descent.

"Easy boy. Easy. We're doin' rightly. Take it nice and cushy on the way down. Good boy." Man and beast worked in total understanding – the perfectly cut sward to their left was testament to that. Hughie talked ceaselessly to the powerful Clydesdale each time they worked together. It was something he had learned from his father, a way of concentrating on the task in hand. His father had bought Major at Carrickmacross Fair shortly before taking ill with the influenza that ultimately killed him. Major was therefore a bond between father and son and provided a continuity on the farm. Talking to the horse also helped prevent Hughie's mind from reflecting too much on the recent past. Earlier in the day he had let his mind wander and, on descending the Hill Field, he imagined he saw Sarah standing on the bottom bar of the five-bar gate waving cheerily to him. He checked himself before returning her wave, rested briefly on the shafts of the plough and said a prayer for his sister.

The plough jarred suddenly as it hit a rock, sending a violent shudder through Hughie's arms. "Whoa, Major!" he called. He set about dislodging the stone, digging out the soil around it with his hands.

"God bless the work!" The voice was strange, the accent even stranger. Hughie looked up to see a burly uniformed figure towering above him. It was a member of the newly formed Civic Guard, An Garda Síochána.

"Work like this I could do without," Hughie replied, staying on his knees.

"Here, I'll give you a hand." The guard dropped on his knees beside him. Together they levered the triangular rock out of the soil and tumbled it out of the path of the plough.

"Thanks!" Hughie panted.

"No bother. We have rocks like that in Kerry. Brosnan's the name. I'm based in Carrick."

"You're a long way from home."

"I am that." There was an uneasy pause.

"That's a fine animal you have there."

"The best. He's paid for," Hughie added quickly.

"Is that what you thought I'm here for?" The man laughed. "I'm only making a few inquiries. There's been a bit of IRA activity around here lately. I'm wondering if you noticed anything unusual in the last while."

"Not a thing. It's a quiet part of the world."

"Mmm. And you'd be?"

"McKevitt, Hughie."

"From below?" He nodded towards the house.

"Aye."

The guard took a pencil and notebook from his breast pocket. "And who else would be living – ?"

"Brothers James and Pete. Sister Bridie."

"That's all?" He noted the names in his notebook.

"That's all."

"It's just for the record, you understand. You'll keep an eye out for anything –?"

"Hup, Major!" Hughie called out. The plough cleaved into the soil once more. Guard Brosnan smiled and brushed the soil from the knees of his trousers. He sauntered back towards the gate where he had left his bike.

"Good luck with the crop," he called. "Barley I suppose?"

"Oats."

The Kiltyrea Band Hall was a squat, drab, flat-roofed building set in a small grove of pine trees. It was built originally as a practice hall for the local fife and drum band but after the war it became a Protestant men's drinking club. Winston Pollock frequented it regularly and, in recent months, he had introduced a new member to the club.

"We'll have one more for the road, Gordon. There's a nip in the air that demands a Black Bush to heat us up, eh?"

His partner nodded silently.

"We'll need to keep an eye out on the way home. Them IRA fellows are causin' some bother round here lately – but we're ready for them – aren't we, Gordy?" He patted a bulge in his jacket just above his hip.

"Rose will be cross." Gordon spoke like a child who was fearful of a scolding for coming home late.

"You leave Rose to me," Winston said, putting a hand on Gordon's shoulder. "I know how to handle Rose. She's a good woman." The barman put two whiskeys in front of the men.

"A good woman. A good woman." Gordon echoed his friend's words dreamily, then roused himself from his reverie. "We'll have the same again, Sammy," he called loudly to the barman.

In the event Winston did not confront Rose. He dropped Gordon at the gate of Sloe Hill. Then he suddenly remembered that he had promised to call on his brother-in-law. Rose was in the pantry when Gordon stumbled on the step at the back door and almost fell into the kitchen.

"A fine condition to come home in," she said coldly as she draped muslin over the milk jug. "And a fine time too."

"Don't be cross, Rose," her brother whispered.

"I have every right to be cross. And more than cross." She wiped her hands anxiously on her apron. "Things can't go on like this. They just can't. Mother is getting more demanding. The babies are fractious. I think Hannah is sickening with something. I've only just got her asleep. The stock weren't fed this evening. And you tell me not to be cross!" She began to sob quietly into her apron. Gordon slumped into a chair and buried his head in his hands.

Rose tried to compose herself and began to tidy things away from the table.

"I know . . . I know you're grieving, Gordon." She spoke haltingly, deliberately rattling the crockery to drown the pauses in her speech. "But we can't go on like this. I can't manage everything." She sat on a chair opposite her brother.

"Bridie McKevitt will be down tomorrow. I suggest we put a proposition to her."

Her brother began to shake his head violently without removing his hands.

"Hear me out, Gordon. Hear me out."

Rose spoke firmly for a full ten minutes.

There was an occasional sigh but no other reaction from her brother. When she had finished she stood up and took the oil lamp from the mantelpiece.

"I'm off to my bed, although I doubt if Mother will give me much peace. There's some stew in the pot that I kept for you. You could do with it. Goodnight!"

Dark invaded the room as she left the kitchen. An occasional tongue of flame leapt from the dying fire, casting weird shadows on the kitchen walls. The crouched figure at the table slid slowly downwards until his head rested on the table itself. Throughout the house clocks began to chime midnight.

Bridie McKevitt dismounted from her bicycle and wheeled it up Caraher's Brae. She had not gone twenty yards when a familiar dishevelled figure clambered over a gate and approached her. Jemmy

Patchy's presence lifted her. There was a time when she would have considered him a mild nuisance but now he was one of the few constants in a fast-changing world. He was a link with Sarah who had cared for him more than anyone. Bridie knew Jemmy grieved for Sarah. She had met him only a few times since Sarah's death and each time he was morose and sullen. Today however he was quite agitated.

"Permission to speak, sir?" he blurted before he even reached Bridie.

"Of course, Jemmy."

His eyes looked furtively about him.

"Bulgars, sir. They have infiltrated our line. We've had several sightings." He stood in front of Bridie. There was genuine pleading in his eyes. Bridie felt uneasy. Sarah had been much better at handling situations like this.

"I – I'll have it investigated, Jemmy." She remembered something which would change the subject. "Call down to the house. I think James has a pair of boots for you."

He saluted her and stood aside to let her pass.

Rose Johnston sat rocking the double pram gently by the kitchen fire.

"I'll come straight to the point, Bridie. Things have not been easy for us since Sarah died." She whispered the last word. "You know that. Gordon is – grieving, sorely." Bridie understood the hesitance in Rose's voice. "Mother is growing more demanding. And this pair," – she managed a wan smile – "would take all my time if I had it to give them! I've talked with Gordon and we've decided to ask you – if you'd take one of them – for a while – until things improve . . . "

Her voice was breaking up. Bridie noticed Rose's knuckles whiten as she gripped the bar of the pram fiercely.

She was taken by surprise with Rose's offer but had no hesitation in accepting it. She owed it to Sarah, if not to the Johnstons.

"Your brothers won't mind?"

"It'll learn them well! They'll probably spoil him rotten anyway!"

"You said him – you'd prefer –?"

"Och, that was just a slip of the tongue!" She paused, then added, "But I think I'd like to have Brian anyway." She didn't understand

93

why she said that. The words fell from her lips.

"It will be done. I'll tell Gordon," Rose sighed. To their mutual embarrassment, they both said "Thanks" simultaneously.

Three days later Winston Pollock drove Rose to McKevitts'.

"Gordon – couldn't come," she said as she handed the baby to Bridie. "There's just a few bits of clothes – and his bottle. That's all. What about a pram? What will – "

"It's all right. Owen Brady has got us the loan of one. Will you come in for a cup of tea?"

"No. No thanks. I have to get back to Mother – and Hannah."

She bent over the sleeping form.

"Take good care of him. Goodbye." She turned away quickly and slipped into the car which sped away instantly, narrowly missing Owen Brady as he turned into McKevitts' yard.

116 Poplar Ave
Philadelphia

Dear Bridie,

Jesus, Mary and Joseph, Bridie. My hand is shaking as I write this. Your letter is on the table before me but I can't believe it. I won't believe it. Dear lovely Sarah. Why, Bridie, why?

I couldn't go to work today. I just cried and cried. Sarah. Lovely Sarah. Gone from us. And the little ones left. Sarah. I can see her laughing at the dance. Always laughing. Oh dear Jesus, Bridie, what are we to do? I'm not much good at praying but I'll try. This is a total waste of a letter. I'll not go on with it.

God bless,

Jimmy.

CHAPTER THIRTEEN

Podgy Kernan reined in his pony and trap as they entered McKevitts' gateway. He squeezed past the contents of the trap with some difficulty and tied the pony to the gatepost. He wanted to conceal the surprise until the last minute.

"God bless the work!" he called as he entered the kitchen. Bridie was leaning over a scrubbing board which stood in a large enamel basin of soapy water. The voice startled her.

"God, you put the heart crossways in me!" She gulped. "Where did you come out of? I heard no noise – "

"I didn't drive all the way in. Are ye not pleased to see me?" He wiped beads of sweat from his forehead with his shirt-sleeve.

"Well it's seldom enough we see you round here. Is there something wrong?" She resumed scrubbing.

"Divil a thing. How is things going with the baby?"

She held up the nappy she was scrubbing.

"It's a full-time job, I can tell you. I think the men of the house are getting a bit jealous. Too well looked after they were up to now."

"Any man that has you looking after him is doing rightly." Podgy beamed in anticipation of opening up a particular line of conversation. Bridie replied by scrubbing the nappy even harder.

"Have ye no help at all?" Podgy asked, somewhat deflated.

"Josie's here a couple a times a week."

"That'll be good trainin' for her surely."

Bridie piled up the nappies on the scrubbing board and brushed past Podgy on her way to the wringer outside the back door.

"Hold on! Hold on!" He hobbled after her. "I have a surprise for ye!"

Bridie looked at him with some puzzlement. He hurried past her and beckoned her to follow him. She laid the washboard on the ground and followed him around the gable end. He reached the trap and proudly drew back the grey blanket that covered his surprise.

"What do ye think of that?"

It was a cot, gleaming white in the morning sunshine.

Bridie was genuinely surprised.

"It's – lovely – but where, how – ?"

"There's a young Hughes fellow comes into the bar. He can do anything with wood. D'ye know what that cot is made of?"

Bridie shrugged her shoulders.

"Butter boxes! As true as God! You wouldn't know it in a million years. Isn't it a grand piece of work?"

"Grand. But how much – "

"Don't be talking about that. It's my gift to you, a token of – well, you know what I think of you, Bridie."

There was an embarrassed silence.

"It's a lovely thought, Podgy, but I couldn't – "

"You could and you will and that's the end of it. Except a cup of tea and a few cuts of your soda bread wouldn't go astray!"

"You're a terror, Podgy Kernan." Bridie helped him unload the cot from the trap.

"At least I can bring a smile to your face. There's hope for me yet."

Hughie joined them for tea.

"Begod, you're your mother's daughter, Bridie McKevitt," Podgy mumbled through a mouthful of soda bread. "We used to say no one could make soda bread like her, but yours is as good any day. D'ye know, I forgot to tell you – I was stopped by the Guards on the way here! They thought I had a crate of rifles with me! Them's the boys got a land, I can tell you!"

"They're around here a lot lately," Hughie said.

"Looking for smugglers, I suppose."

"Looking for IRA men. Them fellows don't recognise any border."

"Well, they're stuck with it now," Podgy said, as he reached for the last slice of soda bread.

"Try tellin' that to the likes of Owen Brady," Bridie said, emerging from her room with the baby.

"Ah, that fella'd be better off delivering the letters on time than guffin' on about a thirty-two-county state," Podgy replied angrily.

"Well, is this the man of the house?" he added in a softer voice. He went over to the infant. "Aren't you the lucky man with a new mammy to look after you? And a good Catholic home to grow up in?" The baby made strange and began to bawl.

Hughie looked at Bridie. "You have a point there, Podgy," he said.

James came into the kitchen. "Good man, Podgy. Any news?"

"Smelled the tea again," Bridie laughed. "You only have to open the tea-caddy and he's down from the loft before it's drawn." She showed Podgy's present to her brother.

"That's a grand bit of work all right," James said, examining the joinery with interest. "It's right good of you, Podgy."

"Och, it's only a small thing. You've had your share of bad blows in this house."

James lit a cigarette and drew deeply on it. "Well, any news?"

"I hear Pat Trainor is home. He signed Aggie into the County Home."

"God love the creature," Bridie said as she poured James's tea. "Sure she never harmed anyone."

"The woman was a heartscald to me anyway," Podgy said, throwing the dregs of his tea into the fire. "She's totally gone in the head."

Later that week the McKevitt family had another visitor. They had just finished the rosary when the black-suited figure tapped at the window.

Pete opened the door.

"Begod! Pat Trainor, I do declare! I'd hardly know you in that get-up! Come in! Come in!"

The family greeted him in turn.

"Pull up a chair. Are ye home for good?" Hughie asked.

"Not at all. I only got a few days – to settle up some family business."

"Will ye take a bit of porridge?" Bridie asked coldly from a dark corner of the kitchen.

"No thanks, Bridie." There was a tremor in his voice before he suddenly burst into tears. The other men shuffled their feet awkwardly and looked away.

"I'm sorry," Pat Trainor sobbed. "It's just that the last time I sat at this fire – Sarah was here beside me . . . Such a lovely girl . . . I cried and cried when I heard the news . . . "

James lit a cigarette from an ember of the fire.

"Ye'll take the porridge anyway," Bridie said, thrusting the bowl in front of him. He accepted it and began eating hurriedly.

"Are ye playing any football, Pat?" Pete asked between mouthfuls.

"Divil a bit. There's only one other Irish fellow in the seminary. The rest of them play ould soccer football. I have no time for it."

"God, we could do with ye back here at the minute," Pete sighed. "We have a half-forward line that wouldn't score if the goals were the width of the field. I mind the time you scored them three points with the left foot against Mullaghbawn! We had a decent team those days."

The conversation drifted on about football and local happenings until Pat Trainor rose to go at midnight.

"The weans. How are the weans?" he said suddenly to Bridie.

"They're doing rightly. We have the boy here."

Pat was visibly taken aback.

"I didn't know that. Could I see him?"

"Well, he's asleep at the minute – "

"I'll not disturb him."

Bridie took the lamp and led him into the bedroom. Pat leaned over the cot and stroked the infant's head with one finger. "He's a fine child, God bless him." He reached into his pocket and withdrew a half-crown. He slid the coin under the infant's hand.

"May he always have health, wealth – and luck." There was a sudden intake of breath as Pat Trainor stood upright and offered his hand to Bridie. "I must go now. I've overstayed my welcome surely." He made his way quickly through the kitchen, calling in a low voice to the three figures hunched around the fire, "Good night lads. I'll see yis sometime."

The black Ford car jerked to a halt at the entrance to Kiltyrea Band Hall before shuddering in a wide arc as it turned right onto the main road. Its sole occupant was crouched over the steering wheel, the car's erratic path betraying his inebriated state. The car brushed against the high grass verge before lurching violently to the right when the driver failed to slow down at a sharp bend. He righted the car with some difficulty and was then confronted by a large tree bough which lay across the road. The driver braked hard and the car slewed right up to the branches of the bough before it came to a halt. Two shots rang out, shattering the windscreen and the driver's brain. The driver slumped against the door. His body fell slowly out of the door through a tangle of branches. Two more shots whistled through the foliage and thudded into his lifeless body. There was a wild whoop from the hazel thicket and then silence, save for the ticking of the car engine.

Winston Pollock stood at the gable end of the Band Hall, relieving himself. He was worried. He had left Gordon Johnston, who had been feeling unwell, sitting in his car. Then shots rang out – their echo in the still night tautened every muscle in Winston's body.

"Jesus Christ!" he roared. "They've done Gordon!" He ran blindly in the direction of the shots.

Pete McKevitt was woken by the chiming of the clock on the kitchen mantelpiece. Two o'clock. He would take another look at the roan cow. She was on her first calf. As he entered the byre, he could hear her restless movement in the stall. He lit the oil lamp. The cow gave a low moan and grew more agitated. "Nothin doin' yet, girl?" Pete talked gently as he stroked the cow's rump. He had a particular affection for this animal. It had been one of twin calves. Sarah had owned the other one. He stood there whispering to the cow, reassuring her.

Shots rang out. Two more. Pete suddenly felt cold. "Bad business afoot tonight," he said. He stayed with the cow for another fifteen minutes. "I'll make a mug of tea and I'll come back to you." The night grew chilly. As he made his way across the haggard, two figures sped furiously down Flaxpool Lane on their bicycles. Their speed and sudden appearance startled Pete. He could only catch a brief glimpse of them, but one of the cyclists seemed to have the distinctive gait of Owen Brady.

James and Bridie called to Sloe Hill to pay their respects to Gordon Johnston. Apart from Rose and Winston Pollock, the many others who passed silently through the house were strangers to the

McKevitts. Rose greeted them formally and invited them into the drawing-room.

"We'll be grand here in the kitchen," Bridie insisted. "We'll not stay long. We just wanted to say how sorry we are. And if there's any way we can help out – with the funeral, cooking, anything . . . "

"Thank you, no," Rose whispered. "The neighbours have been very good." She poured a cup of tea for each of them. "Very good," she repeated.

"And why wouldn't they be?" Winston Pollock growled from the fireside. "The decentest man that ever lived."

"I have to go to Mother," Rose said abruptly. "You'll excuse me."

"Of course." Bridie rose from her chair. "We'll be going anyway," she said, glancing at James.

"The decentest man that ever lived," Winston repeated, his voice trembling. He buried his head in his hands.

"Of course it was me they were after. That poor unfortunate man – " His voice broke into a sob. He took a long drink from the glass of whiskey that sat on the hob. "But they'll pay for it. The bloody bastards'll pay for it. We'll see to that!"

Two gardaí called to the McKevitt household a couple of days later. "We're just making a few inquiries about the shooting," the senior garda explained. "In case you saw or heard anything unusual . . . "

"He was our brother-in-law," Bridie said calmly. "We are all very shocked."

"I'm sorry." The garda was quite taken aback. "I didn't realise – "

"And we saw nothin'," Pete affirmed.

The next day the gardaí came across Jemmy Patchy standing guard outside the derelict house where he slept. He saluted the gardaí.

"Permission to speak, sir?"

"Aye," the bemused garda replied. He gave a knowing nod to his young colleague.

"I have captured a consignment of Bulgar rifles – sir!" Jemmy announced proudly.

"Oh aye. How did you manage that?"

"Pincer movement, sir. They had secreted the rifles here but I came around the west side while General Brodie took the east. Element of surprise. Very important, sir."

"Indeed." The young garda barely stifled a giggle.

"And where would the rifles be now, Colonel?" the senior garda asked with unconcealed cynicism.

"Right here, sir. Awaiting your inspection." He led them over to the ditch and beat back the tall grass. Two rifles lay at the bottom of the ditch.

On the following Sunday afternoon, Owney McCabe took his dog rabbit-hunting through Coomataggart. Owney studied the rabbit runs before deciding where to peg his snares. As he hammered a peg into the ground, he heard his dog give a series of low growls. He laid aside the snare and went in search of the dog. He found him crouched on all fours in front of the Mass Rock. The body of Jemmy Patchy lay sprawled across the rock. His throat was slashed open. Beside him, crudely daubed on the flat rock in Jemmy's blood, were the words ULSTER FOREVER.

CHAPTER FOURTEEN

THE CHURCHYARD IN CULLYBOE BRIMMED WITH NOISE AND ACTIVITY IN THE gloom of an October evening. On one side the gloom was dispelled by the tilly lamps that hung in the mission stall. A group of women pressed in on the stall examining rosary beads, statues and holy pictures. On the other side, their children played hide-and-seek among the shadowy yew trees. Outside on the street men gathered in knots of two and three and engaged in hushed conversation, their faces illuminated occasionally by a long draw on a cigarette.

All was changed at precisely five minutes to eight when Canon McKenna emerged from the presbytery with the missioner, Father Benignus, a portly red-faced man whose considerable girth was amplified by a large leather belt, tightly drawn below the waist.

Canon McKenna stood at the gate. "Now men, now men," he snapped impatiently.

The men shuffled obediently into the yard topping cigarette butts into the gravel as they went. As the men approached, the women scurried towards the church calling the children who had already been frightened out of the trees by the clapping hands of the dark figure who moved threateningly through the yew grove. An eerie quiet descended on the yard. The men straggled into the church porch.

"I'll tell yis something," a voice rose above the whispers. "Youse boys are in for a roastin' tonight."

Father Benignus hitched up his belt and leaned forward in the pulpit.

"My dear people – and how wonderful it is to see the children here. The lovely, lovely children made in the true likeness of God. Innocent and pure. 'Suffer the little children to come unto me,' said the Lord. My dear people, I want to talk to you tonight about a gift. A very precious gift . . . When I was a wee fellow – just like that lovely boy in the front row there – when I was a boy I got a gift one Christmas from my uncle. Do ye know what it was? I'll tell ye. It was a wee train – just an engine and two wee carriages. Simple wee things, painted in red and blue. No track, so you made it go wherever you wanted. Och, was I delighted with this present, my dear people? Hour upon endless hour I played with it. I was the engine-driver and I could go anywhere in my little red and blue train. I could go to Crossmaglen or China. It was the most wonderful thing I ever owned . . .

"Or so I thought. When I was a child, I saw the world through the eyes of a child. But now as a man I see differently. I own a gift far more valuable than the little blue and red train. A priceless gift – hard-won and handed on by loving parents. But I'm not alone . . . "

He paused for a long time, causing the two youths in the last pew who were carefully tying Petesy Marry's bootlaces together to stop and look up guiltily.

"Oh no, I share this wonderful gift with each one of you here tonight from the oldest to the youngest. I speak of course, my dear people, of the gift of FAITH . . . "

The priest's voice echoed through the packed church.

"The faith of our fathers. Aye and of our grandfathers, and their grandfathers going right back to Patrick himself. Belief in a loving God and unflinching loyalty to his commandments. A faith worth living for – even more so worth dying for. And many did. In dark and evil days. Penal days. But the faith survived – as it always will. 'Behold, I will be with you all days,' said the Lord. In this year of 1929 we celebrate the centenary of Catholic Emancipation – the glorious deliverance from penal times won for us by the great Liberator, Daniel O'Connell. Yes, of course we celebrate and we rejoice. But we must be on our guard too . . . "

His voice soared away into another long pause broken only by sporadic coughing. Petesy Marry's bootlaces were now securely tied together.

"There are enemies to that faith within and without. There are new fads being introduced from pagan sources – new music, books, talking pictures. All evil influences, if we do not exercise control over them, my dear people. And in another part of our sundered land, there are those of other faiths. Faiths, yes, but not the one, true faith of the holy, Catholic and apostolic Church we all belong to. So my dear people – and especially dear parents of these lovely, lovely children – we must be on our guard against the world, the flesh and the devil. Cherish that great gift! Now and always!"

Father Benignus suddenly burst into song: "Faith of our Fathers, Holy Faith."

He motioned to the congregation to rise and join with him in the singing. As they did, there was a tremendous clatter in the second last pew as Petesy Marry pitched forward and knocked over the pew in front of him. Above the hymn, his whining voice rang out:

"Ah, Mother of Jesus, which of youse did that?"

The doors of the church porch burst open. Canon McKenna stormed in, his eyes boiling with rage. He made straight for the two miscreants in the back pew, ordered them out of the seat, grabbed each of them by an ear and frog-marched them out of the church.

Hughie McKevitt joined the queue for confession. He was unfortunate to be preceded in the queue by Lizzie Rogan. There were winks and sniggers among the waiting penitents as Lizzie loudly told her life story to the missioner. She was just about to emerge from the confessional when an afterthought occurred to her.

"Oh, and another thing, Father . . . " She disappeared again.

Hughie sighed and resigned himself to a further wait.

"And how long have you been keeping company with this girl?"

"About four years, Father."

"And the passionate kissing. How often?"

"About once a week, Father." Hughie felt his throat go dry.

"And touching? Immodest touching?"

"Now and again, Father," Hughie whispered.

"And has there been full intercourse?"

It seemed to Hughie that the missioner roared out the question to the entertainment of those waiting outside.

"No, Father," came the croaked reply.

"Well that at least is good. You must pray now to Our Lady to keep you out of temptation and you must see your girlfriend in the likeness of the Virgin Mary. Respect her purity always. Do you intend to marry this girl?"

"Yes, Father. Next year."

"Well you must respect her until you are blessed by the holy sacrament of matrimony. Our Lady will help you. Now for your penance say a rosary. Is there anything else troubling you, my son?"

"Well, yes, Father." Hughie began the story of Sarah's marriage hesitantly but with the priest's encouragement he slowly grew more at ease.

"Over the past year they have said that Brian should be returned to them to be with his sister, but we're not too happy. He mightn't get a Catholic upbringing."

"And why wouldn't you be unhappy? The boy – the children – are being deprived of that marvellous gift I spoke of in my sermon. The one true faith of the Catholic Church. Why wouldn't you be worried? Stand firm, my son. That's my advice to you and your family. Hold on to the boy and bring him up in your own tradition. The faith of our fathers. Isn't that what your sister would have wanted?"

"Uncle Petie! Uncle Petie!" The boy ran towards his uncle who immediately swept him in the air and swung him round in a wide arc.

"For God's sake, Pete, will you mind the furniture, not to mention the child," Bridie called from the scullery. "And leave them boots in the porch! I might as well be talking to the wall."

Pete propped the child in front of him and sat opposite, offering him a muddy wellington.

"What did you do to Aunt Bridie today, Brian? You have her in an odious bad mood. Here. Give that oul' boot a pull, will you?"

The child willingly took hold of the boot and strained to ease it off Pete's foot.

"Good man youself!" Pete urged. "She's comin'. She's comin'!"

"Ach for God's sake, Pete!" Bridie intervened. "You have no more sense than the child. And me only after putting a clean jumper on him! Will you go out to the porch and do as I say!"

Pete hopped to the porch with a plaintive sigh. "Some people were just not cut out for motherhood," he muttered.

"What was that?" His sister flashed a suspicious look at him.

"I said . . . eh . . . there's a brave crop of spuds out there," Pete spluttered.

"Hmph! You seem to have brought half of the three-cornered field in with you."

"Uncle Petie, can I go with you to the football?"

"Well there's nothin' surer, little man. I'll expect a full turn-out from this house," he said pointedly. "It's not every day Cullyboe gets to the final of the Feis Cup."

"Feis Cup. Feis Cup," Bridie said. "There's nothing else talked of in this house. Them Drumgrish fellows will walk all over yis."

For all Bridie's protests, there was indeed a full turn-out from the McKevitt household for the Feis Cup final in Castleblayney. Brian sat wide-eyed in the trap beside Bridie as James guided the horse through the town to the football field. Hughie and Josie McLarnon made their own way on bicycles. Bridie feigned indifference to the whole occasion but as soon as the match got under way, she found herself caught up in its intensity.

The rival supporters hurled contumely at the opposing team among shouts of praise for their own players.

"G'outa that, ye bousy!"

"Wouldn't kick snow off a rope . . . "

"Good man, McArdle. Drive it, will ye."

"Go on, Joe McMahon. Ye little daisy."

Pete McKevitt, as captain and centre-half back of the Cullyboe team, led by example and constantly urged on his players. Each time he got the ball, Bridie would lift Brian up to see his uncle.

"Come on, Cully! Come on, Pete!" they would shout.

James remained aloof. He simply nodded or shook his head and smoked cigarette after cigarette. At half-time Cullyboe were two points ahead.

"Will we do it, James? Will we?" Bridie asked excitedly.

"Of course. Haven't we the wind – and a great centre-half back?" He summoned a hawker and bought a stick of "Peggy's leg" for each of them.

The second half was fought with an even greater intensity that bordered at times on ferocity.

"Come on, Drumgrish! Every man a man!"

"Keep him out for Christ's sake. Keep him OUT!"

"Take your point!"

"Ah ref, go home and get your glasses!"

James's forecast was borne out as the strengthening wind helped Cullyboe to stretch their lead. Near the end of the game Pete was felled by an ugly high knee-tackle from his opponent. Bridie could not restrain her anger.

"Boo! Ye dirty little scut!" she roared. A large red-faced man turned to face her, his eyes bulging from his head.

"Ah shaddup, missus. What would you know about it?"

"I know dirt when I see it any day. And when it's Drumgrish dirt, it's the worst of all," Bridie retorted to a chorus of approval from the Cullyboe fans.

The red-faced man was clearly taken aback.

"Ah – Cullyboe for snots and show!" He spat the words out as he turned away from her.

James was riled by his comments. "Any more of that language and the snot will be beaten out of you, friend."

"By you and what army?"

The jeers and taunts of the Cullyboe supporters answered him. To add to his misery, Pete McKevitt recovered from his injury and drove the resulting free all of fify yards over the bar.

"There's yer answer, tomato face!" a young Cullyboe lad shouted. The red-faced man glowered and then moved away to further taunting.

"Why is everyone cross, Aunt Bridie?" Brian asked as he clung to her coat.

"Ach, it's just football, child." Further explanation was drowned out by the din of the Cullyboe supporters greeting the final whistle.

"We won! We won!" Bridie sang as she danced with Brian on the grassy mound already vacated by the fans who swarmed onto the field. "Up Cullyboe! Come on, Brian. We have to see Pete get the cup!"

Pete McKevitt was carried shoulder-high across to the table at the side-line where the chairman of the County Board presented him with the Feis Cup. The supporters roared as Pete held the cup aloft. They pressed forward, eager to touch the cup or slap a player on the back. Pete climbed onto the table and silenced the crowd with an imperious gesture. He fought momentarily for breath, wincing from the pain in his back.

"This is a great day for Cullyboe!"

Roars of approval.

"We never won anything before. People didn't even know where Cullyboe is . . . "

"Well begod they know now, especially the Drumgrish fellows!" The unmistakable voice of Podgy Kernan interrupted Pete.

"Free drinks for all in Podgy's tonight!" a young voice called, to the delight of all.

"There'll be a bottle of stout for each of our heroes," Podgy retorted. "The rest of youse blackguards can pay for your own!"

A chorus of boos greeted this remark before attention returned to Pete.

"I just want to congratulate all the lads for the hard work they put in to win this – " He held the cup aloft to another mighty cheer. "And now that we have a bit of silver I say we should start collecting more!"

James perched Brian on his shoulder so that he could see Pete. When Pete caught sight of the boy he jumped from the table and made straight for his family. He inverted the cup and placed it helmet-like on Brian's head.

"Up Cullyboe!" Brian shouted.

"That's the man." Pete laughed.

Hughie and Josie joined the McKevitts. "Well done, Pete," Hughie said, slapping his brother across the shoulders. Pete winced again.

"Are you all right?" Bridie asked with some concern.

"Just a bit sore. Nothin' a bottle of Podgy's stout won't cure! That fellow was lucky it was near the end of the match or I'd have creamed him!"

"You should have heard your sister stand up for you!" James interjected. "Nearly started a row, she did!"

"Good on ye, Bridie. I always knew you were my best supporter!"

"Can I get some sweeties, Uncle James?" Brian asked.

"You'll get your sweeties and more," James chuckled, "because the McKevitt family are going to have high tea in the Bon Bon restaurant and I'm paying the bill!"

"Oh, James. The Bon Bon!" Bridie queried. "That'll cost you a fortune."

"Indeed and it won't – but it'll cost the next poor divil who wants a bit of saddlery done a fortune!"

They all laughed.

"Look at us. All the McKevitts, all laughing. When did you last see that? We've had our share of bad days. This is one of our few good days. Come on and we'll enjoy it!"

Bridie wiped a tear from her eye and, in a rare show of affection, she kissed her brother. It reminded James of how Sarah had greeted him the night he came home from Dublin.

CHAPTER FIFTEEN

53 West St.
Philadelphia
USA
Jan 30 1930

Dear Bridie,

First of all my apologies for not writing to you for Christmas but things have not been great here, as you might know. Don't ask me to explain the depression, as they call it. All I know is like many others I have lost my job. Just told to go. No more work. It didn't make for much of a Christmas I can tell you. Here I was with my soup and a scrawny chicken that was as old as Petesy Marry's dog and thinking of the goose and roast spuds at home . . .

As you can see I have moved out of my lodgings. Couldn't afford them any more. There's six of us sharing a room here on West Street now. Not exactly the lap of luxury! And there's a Westmeath fellow with a snore that would wake the dead. We go out every day to search for work but so it seems does everyone else. To cap it all it's murderous cold here at the minute.

We just hope things will get better when the spring comes. Maybe you'll say a wee prayer now and then. I can survive as I had some money saved – believe it or not for a trip home this summer. I was going to surprise you all – and put Podgy in his place! Don't give up on me yet – you'd never know.

The one thing that cheered me up was the news of Cullyboe's great win. I never thought they'd win anything without me! Give my congratulations to Pete. You don't know what it means to me to have that little cutting from the *Democrat*, to see all those names I know so well. I take it out and read it every day. It's like a prayer.

Give my regards to all the lads, even Podgy, the ould get! Write to me here if you can. I don't know how long I'll be here, but try anyway. Whatever you do, don't forget me.

Always,

Jimmy.

Bridie put the letter behind the Sacred Heart statue on the mantelpiece. James noticed her dabbing her eyes with the fringe of her apron.

"Not good news?"

"No. Times are bad over there. He lost his job. And there's not much hope of anything else."

"I don't suppose he can get home?"

"No." Bridie began washing potatoes for the dinner. "He sends his regards to you all, especially Pete. He was thrilled with Cullyboe's win."

As she spoke, Pete crossed the haggard towards the cowshed, carrying a shovel on his shoulder. Bridie watched his slow gait with concern.

"What do you think of Pete lately?"

"How do you mean?"

"D'ye notice anything about him?"

"He's been a bit on the quiet side, now that you mention it."

"I don't think he's well. I think it's since he got that knee in the back from that Drumgrish fellow."

"But that was months ago. Did you say anything to him?"

"Och I tried, but he took the nose off me."

At that moment Brian entered the kitchen.

"Well there's my wee helper at last! Where have you been, Brian? I had no one to help me count the spuds."

The child immediately ran to Bridie's side, clambered up on the chair and began counting as each potato was dropped into the pot.

James got up from the fire to go back to work.

"You see if you can draw him out," Bridie called.

"And one for the pot!" Brian sang out.

"Och but you're the smart wee rascal," Bridie said, gently ruffling his hair.

111

Hughie McKevitt and Josie McLarnon were married in Cullyboe Church a week after Easter. A wedding breakfast for the twenty guests was held in the McLarnon home before the couple would catch the train at Culloville to spend a few days' honeymoon in Dublin.

When the meal ended, Josie was prevailed on to join with her brother Matt in playing a few tunes for the guests. Wisecracks were tossed at Josie by some of the assembled men who leaned against the sideboard clutching bottles of stout.

"Begod, Josie, you'll be changing your tune from now on!"

"Hughie's the man'll resin your bow all right!"

"Man dear, there'll be some fiddlin' in Dublin tonight."

Pete and James found themselves isolated in the hallway looking out at mist-covered fields.

"A good drop of rain would be no harm at all," Pete said, eyeing the grey sky. "The early spuds are cryin' out for it."

"There'll be another mouth to feed now – I hope you put in an extra drill!" James laughed.

"Indeed I did – " Pete's reply was cut short by a sudden bout of coughing. He made several attempts to clear his throat before continuing. "Josie would clear a plate of spuds with the best of us."

James lit a cigarette and observed his brother closely.

"Are you all right?"

"Course I'm all right. What'd be wrong with me?"

"I don't know. You haven't seemed yourself for a while."

"Ach it's just an ould cold that's hangin' on since a wettin' I got in March. Puttin' in the same bloody spuds in fact." Pete was restless. He looked back into the parlour as if he were seeking out someone in particular.

"But even before March you were – Bridie noticed – "

"Ach Bridie. Bridie's always noticing." He turned quickly and went back into the parlour. James shrugged his shoulders and followed his brother.

"Aunt Bridie?"

"Yes, Brian."

"Why are we changing all the rooms?"

"Because Hughie and Josie will be home tomorrow and they'll need this room, so you and me are going to Pete and Hughie's room and Pete is going to the wee room under the loft."

She felt bad about putting Pete into such a small room but sacrifices had to be made. She had given up her own room to the newly-weds. It was spacious, had a matching wardrobe and dressing-table in walnut and, most important, it had a double bed. And, by putting Pete at the far end of the house, she hoped they would all be spared hearing his racking cough at night. Her concern for Pete's health had increased to the point where she had mentioned it to Dr McMahon (Pete had flatly refused to go himself). He had given her a cough bottle for Pete but it had given him little respite. She feared he only took the mixture occasionally. Stubborn, just like his father.

Podgy Kernan rubbed his stubbled face and gave Bridie a mischievous look.

"Well?"

"Well what?"

"How's it goin'?"

"God, Podgy, but you have annoying ways. How's *what* goin'?"

"Ah you know. The settlin' in. The newly-weds."

"Oh. We're managing."

"'Tisn't easy for a woman comin' into a strange house."

"'Tisn't easy for us either."

"That's true. That's true. Still there's ways around that, I suppose."

"Twenty Players for James. Like what?" Bridie checked her list against the goods on the counter.

"You know well, Bridie. There's a big empty house here. It's for you if you say the word – "

"Bacon. You never gave me the bacon. And for God's sake not as salty as the last time. Ye had me running to the well all day."

"At least you'll have more help this time," Podgy replied acidly. He rooted under the counter.

"Where's the curse-of-God bacon knife? Hanratty! That eejit is never around when you want him."

"I saw him out in the bottling store."

"It's all right. I have it." He prised the flitch of bacon off the hook on the beam overhead.

"You should get Jimmy McCabe in to run this place for you,"
Bridie said, half teasing, half in earnest.

"Is he home?" Podgy asked, clearly surprised by her suggestion.

"No, but he's lost his job in America and I'm sure he'd come home
if there was a guaranteed job – "

"Ah, that clown. What would he know about running a business?"
he said in a relieved tone. "Are you still writing to him?"

"Aye."

"I see." He held up the bacon for inspection. "You won't get a
better bit of meat nor that between here and Dundalk."

"I'll let you know." Bridie laughed.

"How is Pete? I heard he wasn't the best." The question unsettled
Bridie.

"He – he – has a cough that'd wake the dead. He says it'll go away
and he won't go near the doctor."

"God, I hope it's not the weakness, Bridie. I heard tell of a family
over Clontibret way. Three of them – "

He realised, too late, that he had upset Bridie. He reached to the
shelf behind him, took down a baby Power and dropped it into her
basket.

"Put that into an egg-flip for him. It'll do him a power of good."

Bridie rode home in a daze. Podgy had said the unspoken word
that she inwardly dreaded. "The weakness." No one ever seemed to
use the proper word. Tuberculosis. TB. It was always "the weakness"
or he's "not strong" or consumption. But never TB.

TB. TB. The whirring spokes of her bicycle seemed to sing it out.
TB. TB. Sacred Heart of Jesus, let it not be true. First Sarah, then
Gordon. Now this. Sacred Heart of Jesus I place all my trust in thee.

Her worst fears were realised on a clammy May afternoon when
Brian ran into the kitchen.

"Auntie Bridie! Auntie Bridie! Pete fell down in the spuds!"

Bridie exchanged fearful glances with Josie.

"Get James!" she called and raced out through the haggard and up
the lane to the three-cornered field. Pete lay on his side between two
drills clutching his right side and fighting to regain his breath between
choking coughs.

"I'm all right," he spluttered.

"For Christ's sake shut up!" Bridie roared at him as she hauled him

114

into a sitting position. She wiped the trickle of blood from the corner of Pete's mouth with potato leaves before Josie came stumbling through the drills holding Brian's hand. An anxious James hobbled behind.

"I'll be – all right. Just – lost me breath," Pete gasped.

"Save your breath till we get you home," Bridie muttered as she motioned to the others to help him to his feet. Between the three of them they somehow managed to support Pete on the painful journey back to the house. Josie was despatched to Cullyboe post office to ring Dr McMahon, while Bridie propped her brother up in bed in an effort to ease his breathing.

"He has pleurisy on both lungs. A bad one, I'm afraid. He'll need lots of rest and good nourishing food – eggs, milk and the like."

James lit a cigarette and drew deeply on it.

"Will he be all right, doctor?" he asked.

Dr McMahon looked out at the boy who was shooing chickens across the yard. He shrugged his shoulders. "There's not many recovering from tuberculosis, James. It's a terrible scourge."

"Where in God's name did he pick that up?" Bridie asked in little more than a whisper.

"'Tis easy picked up. Whenever people congregate. Through the football team more than likely." He paused for a moment to watch the antics of the boy and the chickens. "'Tis easy picked up," he repeated. "Ye'll all have to be careful now. What I would suggest is maybe building a wee chalet in the garden. Pete will need lots of fresh air anyway. That room he's in'll not do him much good. Too confined."

"It'll be done," Bridie said quietly.

The rosary was said in a subdued mood that evening. "The third joyful mystery, the nativity of Our Lord," Bridie intoned. There was silence.

"Hughie!" There was no response.

Hughie knelt at a chair, his head buried in his hands.

"Hughie, it's your decade!" Bridie called sharply. He stared blankly at her with reddened eyes and gave out a decade that was punctuated with angry sighs.

CHAPTER SIXTEEN

Podgy Kernan came to the aid of the McKevitt family in providing
the chalet for Pete. He contacted Martin Hughes who had made
the cot for Brian and instructed him to take down an old
summerhouse in his overgrown garden and re-assemble it in
McKevitt's orchard.

"There's the best of timber in it," he assured Bridie. "All it needs is
a couple of coats of paint and she's made for the job."

"You're very good to us, Podgy."

"No bother. Sure, what use would I have for a summerhouse!
Mother put that up one time when she had notions of grandeur. It
was hardly ever used. I used to keep barrels of stout in it myself
latterly."

"Still, it's very kind of you."

"I didn't tell young Hughes what youse want it for. He's not from
round here. You know the way people'd be about the weakness.
Some people," he added quickly.

"I know it too well. And James knows it too. His business has
fallen off. People are afraid to come near the place. Most of them that
do come drop off a saddle or britching at the gate and collect it later.
You'd think we had the plague."

"It is the plague for many."

Bridie painted the chalet herself with the help of Brian who
enjoyed painting in long white streaks. Bridie always made sure to
come after Brian so that the wood got an even coating of paint.

"Aunt Bridie?"

"Mmm."

"Why does Petie have to live out here?"

"It's to help his sickness."

"Can I live out here with Petie?"

"No, love."

"Why?"

"Because Petie has wee bugs in him that you might pick up."

"There's wee bugs out here too – in the grass."

"There's no harm in them." Bridie laughed.

When the painting was finished, Bridie was quite proud of her work. The chalet stood gleaming in the June sunshine, stark amid the rich green of the apple trees. If only it were just a summerhouse, she thought.

Hughie took a deep breath and marched into Pete's room.

"Come on boyo. Hop out there. You're off on your holidays – to your summer residence."

"Oh aye. Some bloody holidays," Pete muttered. He stood wrapped in a blanket while Hughie disassembled his bed.

"They say the Pope has one – a summer residence – so you're in good company."

"He hasn't got this bloody ailment, though, has he?"

There was emptiness in the house that evening. As they rose from saying the rosary, Hughie looked across at the little white house, ghostly in the lengthening midsummer shadows.

"It's not right. It's not right that he should be out there like a – like a leper!"

"No one's happy about it but the doctor says it's his best chance," Bridie replied.

"What if he takes bad during the night – with a fit of coughing or something?" Josie asked.

Bridie stirred the warming porridge vigorously.

"The best I can do there is give him an old biscuit tin and a wooden spoon. If anyone has a better idea, I'll be glad to hear of it. Now who'll take the porridge out to him?"

"I will," Josie said.

"Good! And make sure he takes plenty of milk with it!"

As the summer days slipped by the McKevitt household gradually

117

adapted to Pete's isolation and tried as best they could to fit him into the pattern of their day. Bridie would *"céilí"* with him when collecting eggs or hanging out the washing. James came down from the loft to smoke a couple of cigarettes and share the news he had gleaned from the *Democrat* or from his customers. Hughie would take his dinner out to the orchard and discuss farming matters. He made little of the fact that his workload had practically doubled, but Josie was a good worker and the neighbours helped out. Brian was forever coming over from the haggard with excited cries of "Look what I found, Uncle Petie!" and "Look what I did!" Bridie was concerned for both of them – for Pete whose rest was disturbed and for Brian whose exuberance often brought him into too close contact with his uncle.

Josie played her part. On long summer evenings when the chores of house and farm were done, she would take a chair and her fiddle and seat herself under an apple tree. As the rays of the lowering sun filtered through the branches above, Josie practised her music, mainly slow airs with occasional polkas and reels. In the stillness, the music wafted across the countryside, catching the ears of haymakers who took time out to lean on a fork or light a cigarette and savour the balm of a wistful tune at the end of a long day. Each time the recital would end with the same tune, for each evening Pete would have the same request.

"Give us 'Teddy O'Neill' before you go."

June 29th, the feast of Saints Peter and Paul, was celebrated with the first cooking of Pete's early potatoes.

"Balls of flour," James commented. "God bless the hand that planted them."

"I think he's gettin' better," Hughie declared. "He looks younger, almost. I know he's pale but why wouldn't he be, stuck in that place all summer. He just looks, I don't know, better-looking than he ever did. More good-looking I mean."

"Yeah. I noticed that too," Josie said eagerly. "I told him the other day he looked like a big china doll!"

Bridie said nothing. The doctor had told her this was often a sign of the disease being in an advanced state but she hadn't the heart to shatter their optimism.

Podgy Kernan arrived in pony and trap – a welcome distraction.

118

He wiped his brow with the rolled-up sleeve of his shirt and waved cheerily in the direction of the chalet.

"Good man Pete! How is things in the White House? Begod President Hoover is only trottin' after ye. I hear you're comin' on by the new time!"

There was a muffled response from the chalet.

"Ye never said a truer word!" Podgy bellowed. He took a brown paper bag from the trap and turned and limped into the kitchen.

"God bless all here!" he announced.

"And you Podgy. What ails ye, man?" Hughie enquired.

"Ach this curse-of-God foot is givin' me hell's torment."

"I told you years ago to see the doctor," Bridie reprimanded him.

"Ach butchers and quacks, the lot of them. People dyin' like flies and they can't . . . " His voice trailed away as he realised his foolishness. "Anyway! Anyway! I've been on a little business trip —" he began, as he delved into the bag.

"Sounds like the unapproved road to Newry," James said wryly.

"Now you, cobbler, stick to your last and don't be interruptin' me! I have some presents here for the lady — for the ladies — of the house!"

He proudly withdrew a two-pound pot of jam to the delight of all. "And this is specially for the patient!" He produced a gleaming tin of Tate and Lyle's Golden Syrup. "That's the stuff'll put him right. And this, young man, is for you!" He handed Brian a large bar of Cadbury's chocolate. The boy's eyes lit up.

"What do you say, Brian?" Bridie reminded him.

"Thanks, Mr Podgy," Brian said as he tore off the wrapper.

"It's like Christmas Day in June." Josie laughed.

"Indeed. God forgive you for smuggling on a holy day, Podgy Kernan," Bridie jibed.

"Sure a man has to make a livin'," Podgy argued. "Anyway it's not a holy day in the North — not with the Prods anyway."

"Will ye take a drop of tea?" Bridie said in a kindlier voice.

"Never been known to refuse it. Good and strong, if you don't mind."

"Any trouble with the customs?" James asked.

"Never saw one of them. Must be all at Mass! The only trouble I had was with young Hanratty. Lord save us, but he's the two ends of an eejit. We were bringin' five bonhams over and I told him to give them a drop of stout – just to keep them quiet on the journey. The eejit must have given them a firkin each! They were all fast asleep when I got to Newry. Had a woeful job waking them up."

The laughter rang through the kitchen.

"God, Podgy, but you're the man for the crack!" Hughie chuckled. "Any more news?"

"Aye. I heard old Mrs Johnston – Gordon's mother – died yesterday. Must have been ninety if she was a day. Funeral arrangements later, as they say."

The McKevitts would not be attending the funeral but Bridie felt it right that she should send a note of sympathy to Rose Johnston on behalf of the McKevitt family. She felt sure that her mother's death would be a relief to Rose who could now get on with her own life. Since Gordon's death there had been little contact between the two families, until the past year. Bridie had met Rose accidentally while shopping in Newry. It was an awkward, difficult meeting but what surprised Bridie was Rose's insistence that it was time that Brian "came home to join his sister." Bridie dismissed the idea until a letter arrived a month later. It was brief, formal, almost legalistic in tone.

Dear Miss McKevitt,

Further to our recent conversation, I wish to express once again my conviction that Brian Johnston should be returned to his natural home as soon as possible. I would appreciate therefore if you could expedite this matter.

Yours sincerely,
Rose Johnston.

The letter alarmed Bridie. It angered her brothers when she read it to them one night after the rosary. They decided to ignore the letter and for some reason there was no further communication from Rose. Bridie suspected that her mother's condition had deteriorated but she was relieved when nothing happened. Hughie relayed the missioner's advice to the family and they felt reassured. And then Pete's illness clouded everything.

Bridie was taking clothes from the line when Pete called to her.

"Who's that coming down the lane?"

He had positioned himself in the white house so that he could observe the comings and goings along the laneway. It helped relieve the tedium of long summer days. Bridie strained to identify the erect figure moving slowly along Flaxpool Lane.

"Mother of God! It's Rose Johnston!"

She swept up the basket of clothes, leaving some on the line and scurried back to the house.

"Rose Johnston's on her way! Is the parlour tidy?" she called to Josie as she swept through the kitchen grabbing a pair of boots and slinging them into a corner of the scullery.

"I think so."

"Well go see!"

James sat by a smouldering fire reading a paper.

"James, sweep up them ashes for God's sake!" His reaction was to light a cigarette.

"She'll take us as she finds us," he said calmly. Bridie flashed a look of annoyance at him and tidied the hearth herself. She lifted the porridge pot off the crook and took it to the scullery. She was removing her apron as Rose Johnston knocked timidly on the door.

"Good evening, Rose. You're welcome," Bridie said. "Come in. Come in."

"Thank you. I'll not stay long."

"Sure what hurry would be on you. You know James." He rose awkwardly and nodded to the visitor.

"Hello, Rose."

"And this is Josie, Hughie's wife."

Josie shook hands with Rose.

"Pleased to meet you."

"We were sorry to hear about – " Bridie began.

"Thank you for your note. It was a kind thought," Rose interjected in a courteous tone.

"Well what are we all standing here for! Come into the parlour. Josie will put the kettle on for a cup of tea."

"I'll not stay that long. Don't be troubling yourself. It's just some family business."

Bridie led the visitor into the parlour, a room the McKevitt family used only at Christmas or on rare celebratory occasions. She swept her sleeve along the top of the sideboard as she passed.

"Sit down. Sit down." She nodded to James to join them.

Rose sat on the edge of the armchair.

"I'll come straight to the point. It's about Brian." She paused but neither Bridie nor James spoke.

"I wrote some time ago but circumstances – family circumstances – prevented me from pursuing the matter."

"Brian will be staying here," Bridie said quietly. "It's his home now."

"His natural home is Sloe Hill. I promised his father I would care for his children." Rose looked straight ahead, avoiding eye contact with either of the McKevitts.

"His father promised that both children would be brought up in their mother's faith. Both children." Bridie's voice grew stronger. "There's not much chance of that happening in Sloe Hill."

"The children belong in Sloe Hill. Brian would be the natural heir to Sloe Hill." Rose adjusted her position in the chair.

"That may be, but we have a responsibility to God and the child's mother."

"We are not pagans. We are God-fearing people. The child would – "

"It wouldn't be fair on the child apart from all that, Rose," James broke in. "He's lived here for nearly all of his life. He wouldn't go even if – "

"He'll not be going and that's that," Bridie affirmed.

"It's not right." Rose's voice was beginning to quaver. "I promised his father."

"If right was right," Bridie lowered her voice again, "both children should be down here, being brought up in their mother's faith."

There was a tense silence, punctuated by the clatter of utensils in the kitchen.

"Well I tried," Rose sighed. "I'll go now. I bid you good evening."

"That wasn't very pleasant," James said as he watched Rose cycle back along Flaxpool Lane. Bridie hung the porridge pot on the crook over the fire.

"I wonder how much of it was her idea and how much was Winston Pollock's," she mused. "I hope that's the end of it."

A week later, a furious banging on the biscuit tin brought them all running to the orchard. It was the first time Pete had employed the makeshift alarm. Hughie and Josie raced from the cowshed where they had been doing the evening milking.

"What's wrong?"

Pete fought for his breath. He pointed anxiously to the top of Flaxpool Lane where the sun glinted on the roof of a stationary car.

"It's – Winston Pollock," Pete panted. "And Brian – went up the lane – a few minutes ago!"

Hughie sprinted across the haggard, grabbed his bike from the hayshed and tore up the lane. Bridie and James joined Josie in the orchard and watched and waited anxiously. The sounds of heated argument carried on the evening air, followed by the revving of a car engine. The car shot forward and disappeared from view. Bridie ran to the gate and gave a cry of relief when she saw Hughie return with Brian perched on the crossbar of the bicycle.

She swept the boy into her arms.

"What was he doing, child?"

"He gave me sweets, Auntie Bridie, and he wanted me to go for a drive in his car."

Bridie clasped the boy tightly to her and looked anxiously at Hughie. She squatted in front of Brian.

"You must never take sweets from him, Brian, and never, never go in his car!"

"Is he a bad man, Aunt Bridie?"

"He's a bloody blackguard and if he has any sense he'll not come round here again," Hughie muttered.

CHAPTER SEVENTEEN

W HEN BRIDIE BROUGHT SOME BREAKFAST DOWN TO PETE ON THE
following day, she was struck by his appearance in the
morning light. There was a radiance about him that she hadn't
noticed until now – his complexion was almost translucent. What was
it that Hughie had said about him? - "more good-looking, like a big
china doll."

"How are we this morning?"

"Rightly. Rightly." He eased himself into a sitting position, drawing
the blankets about him. "I could do with another blanket. It was fair
cold the past few nights."

"I'll see to it." As she leaned across to puff up his pillows, she
noticed the blue tinge in his skin. She felt a chill run down her spine.
Dr McMahon had told her to watch out for that. It was another sign
of deterioration. She sat on the bed as he ate in silence. Every few
moments she would steal a glance at him and look away.

Blue. She remembered the blue of the flax in flower. People going
the road would stop to look at it in the long field. It was a
breathtaking sight. The blue faded. She was still looking down the
long field. Fifteen men were ranged across its width, bent to the task
of pulling the flax on a sultry August day. They would start at the far
end and work their way towards the headland. They worked hard
amid banter and laughter about football and girls and the gossip of
the day. And always Pete edged ahead of the others. Hardly eighteen
and stripped to the waist, his lithe bronzed body bent and rose to the
task in an easy rhythm. And though they would jest and tease –

"God, Pete must have a woman waitin' for him at the headland," "He'll not leave a heel of bread for us," – nobody could get ahead of him. He would give a little whoop when he reached the headland where Bridie and Sarah were waiting with cans of tea or buttermilk, great slabs of buttered slices of brown bread wrapped in newspaper and a bucket of hardboiled eggs. The men sat along the headland and ate and drank with relish as they eyed their morning's work with satisfaction. Occasionally someone would give a snatch of a song or a verse of a poem learned at school.

"Many's the time Master Connolly reddened me hands for not knowin' that oul' poem!"

"Ach, he wasn't the worst!"

"He was a cranky oul' get and no mistake! If I had him here now I'd feck him into that flaxpool!"

Laughter. Always laughter. And always the loudest laughter came from Pete McKevitt and Jimmy McCabe.

Jimmy McCabe. No word from him for ages . . . Back to work. Gather the flax, bound in sheaves with dry rushes and carry it to the flaxpool. Pete and Hughie would look at each other in anticipation before they leaped together into the pool where they would sink and stack the sheaves below the water in order to ret the flax.

There were wild shrieks from the brothers as their bodies hit the water and were submerged up to their chests.

"Easy talk to you now, McKevitt!" Jimmy McCabe roared, deliberately eyeing Bridie as she and Sarah made their way back to the house. "It must be blue with the cold down there!" he sniggered.

Blue. The flax in flower. A sea of blue. A tinge of blue . . .

"I said I'll turn into a chicken from eating all these eggs," Pete's voice broke in. "God, I knew I was talkin' to myself. You were miles away, there."

"I was," Bridie smiled. "Miles away!"

Bridie's mind was troubled all through the day. When she had put Brian to bed that evening she slipped out of the house and down the lane. Not wanting to meet anyone, she left the lane and crossed the three-cornered field. On through Barney's field (no one could ever explain how it got its name) and then the steep climb up the hill field. She remembered how, as children, they would spend hours on a

Sunday afternoon doing roly-polys. You lay flat out on the ground at the top of the hill and projected yourself forward, rolling as fast as you could to the bottom. There were arguments and rows about who finished first and who cheated. She shuddered at the thought of rolling down the sun-baked rough surface now, but then it was fun. When you reached the bottom you raced back up to do it again. Now she found herself clutching her side as she fought her way up the hill on a humid evening. She remembered again that the roly-polys were stopped – for a time – when Sarah gashed her head on a protruding stone and a frightened James carried her all the way home.

She sank down in the overgrown grass at the top of the field. She scanned the sweep of the valley below her. Landmarks stood out in the hazy sunshine. Petesy Marry's, Fox's, where three old bachelor brothers hardly ever spoke to each other, Barney McQuillan's – the noisiest house in the parish with six giddy women laughing, shrieking, fighting – McMahons, Quinn's forge, Drominarney schoolhouse . . . She remembered bringing a frightened Sarah by the hand on her first day at school.

Sarah. Sarah. Why did everything come back to Sarah?

What a change had come over this countryside in the past ten years – divisions. A line drawn on a map making two countries out of one. A civil war where brother fought brother. Suspicions, mistrust. Loss. Her parents gone. Sarah and her husband dead. Two orphans causing more division. And now Pete dying. Why did one family have to suffer so much – or did every family suffer? She thought of Brian asleep below. She feared for him – in danger of kidnap, in danger of catching "the weakness" from Pete. A movement below caught her eye. The cows ambled in single file out of the shed and headed for the low meadow. Behind them Hughie and Josie sauntered hand in hand, their occasional laughter echoing in the heavy air. Bridie looked away into the distance beyond the schoolhouse. Through the heat haze she could just discern the outline of a two-storeyed house, flanked by tall pine trees. Podgy Kernan's Bar, Grocery and Lounge. She closed her eyes and made a decision.

Podgy Kernan was struggling with a crate of bottles from the bottling store to the bar.

"Good, Bridie," he panted. "You're a sight for sore eyes!" He dropped the crate with a clatter and wiped sweat from his brow with his sleeve. "Will you have a glass of lemonade? This goddam heat is killing me."

126

"No thanks – but I want to talk to you, in private."

He looked at her quizzically before calling back in the direction of the store. "Hanratty! Look after the shop, will ye!" He ushered Bridie into the kitchen which was as untidy as ever.

"Showp!" Podgy swept the sleeping cat off the top of the range. "Here – sit yourself down!" He cleared the armchair of newspapers.

"I'm fine here." Bridie sat down at the table and cleared a space for her elbows.

"I have a proposition to put to you, Podgy."

"Begod!" he laughed nervously.

Bridie recounted the meeting with Rose Johnston, the incident with Winston Pollock, her growing fears for Brian's safety, her anxiety over Pete's health and the danger it posed for Brian.

"He'll be starting school here beside you in September, so what I want to say to you is this. You've often asked me to come in here as woman of the house. I'm prepared to come in here – with the boy – as cook and housekeeper and nothing more."

She paused to emphasise the last words by drumming her fingers on the table. "In return all we ask is our keep – nothing more, except the freedom to tend to Pete for as long as he lives."

"Begod!" Podgy whispered.

"Is that all you have to say?"

"Begod and it's not. You want to come in here and look after the place – and me?" he added with a broad grin.

"And nothing more. I want to make it clear I'm doing this for the boy. Do you understand?"

"Yes. Of course I do."

"Well?"

"Holy God, Bridie. You know you don't have to ask –"

"I do and I'm asking."

"Of course you can come. You'll be more than welcome. And the boy too," he added. He rubbed his hands at the prospect. "Begod this is the best news I've had for a long time."

"Don't get any wrong notions, Podgy. I'm doing something that will be of benefit to both of us."

"Of course! Of course! When will you be movin' in?"

"As soon as I can. They don't know about this at home yet."

Bridie broached the subject with her family at Sunday dinner. James held a spoonful of stew in front of his face, a face that manifested clear disbelief.

"You're not serious, surely?"

"I am. I've thought a lot about it. It's for Brian's sake."

"But we can manage him here," Josie said. "I hope you don't think – "

"He'll be starting school next month and the school is next door to Podgy's. Neither of us is going to America! We'll be over and back all the time. It's to do with Pete's condition too," Bridie added.

"Podgy must be in his element," Hughie said cynically. "You'll surely set the tongues wagging now."

"They'll wag anyway, with or without me."

"I still think it's not necessary," James said. "This is your home. And Brian's."

"It's necessary, James," Bridie replied quietly. "It's necessary."

She dreaded telling Pete. The tears welled up in his eyes.

"I suppose I'm driving youse out." His voice was raw and hoarse.

"You're doing no such thing. Won't I be up to you every day? It's – just that tangle with Winston Pollock. It makes me nervous. It'll only be for a while, Pete."

"I'm not going to get out of this thing, am I Bridie? It has me bested." He gave a deep racking cough.

"Don't talk like that, Pete. Aren't you eating well? And your colour is good."

"It's not how you look. It's how you feel."

"Whisht now and get a bit of rest."

"Christ, that's all I've been getting for the past months – and I'm tireder than ever."

Bridie made a pretence of looking for eggs in the long field where the hens had wandered.

"Why are you crying, Aunt Bridie?" She dabbed her eyes quickly and ignored the boy's question.

"Where did you come from, you wee rogue?" she smiled.

"I found these!" He proudly displayed an egg in each hand.

"Well aren't you the clever rascal! Where did you find them?"

"Over at the flaxpool."

"You didn't go down the lane, Brian?"

"No. I went the way the hens went. I watched them and watched them."

She ruffled his hair.

"Brian, you and me are going to live in Podgy Kernan's for a while."

"Is that 'cos of the bad man?"

"Well it's mainly because you'll be going to the school down there next month. And because Podgy wants me to help him in the house."

"Do we have to go?"

"Why? Do you not like Podgy?"

"No."

"Why?"

"'Cos his hands are always wet."

Podgy fidgeted with the door handle. "The best room in the house. I sleep downstairs meself. There!" He flung the door open. "Isn't that a grand room? I lit a fire the other night to air it a bit."

The room was spacious and bright. The sombre furnishings would not have been Bridie's choice but that was not an immediate concern.

"Look!" Podgy beckoned them to the window. "You can see nearly the whole parish up the brae. Petesy Marry's. Fox's. Barney McQuillan's – well, yis can hear them anyway! And there's McKevitts! Yis could nearly wave to Pete. Poor Pete. Man dear, it's an awful scourge."

"The room is grand, Podgy. We'll unpack our things," Bridie said hastily.

"Fair enough. I'll put on the kettle for a mug of tea. And I'd say this man wouldn't say no to a lemonade!"

He caressed the boy's cheek. Brian cringed at the touch of Podgy's clammy hand.

In the ensuing weeks Bridie faced into the task of putting order on Podgy's house. The kitchen presented the most daunting challenge. The accumulated dust and grime of years of neglect yielded only to her most determined efforts. She spent a whole day cleaning and blackening the range. Boxes filled with newspapers, empty bottles, jars of moulding jam and badly chipped crockery were hauled out to the dump at the end of Podgy's garden. While removing clutter from the top of the kitchen press Bridie came across a handsomely bound book. She blew the dust off the cover to reveal the title:

The Works of Robert Burns

Intrigued, she turned over the fly leaf. The inscription read:

Christmas 1894
To dear Peter
From Auntie Maud
Beneath the inscription was written in childish hand:
Peter Kernan
Drominarney
Cullyboe
Co. Monaghan
Ireland
The United Kingdom
Europe
The World
The Universe
Age: 12 years

Bridie smiled. She had never known Podgy by any other name. She sat at the table and flicked idly through the book. She recognised the occasional poem – "John Anderson my jo, John," "Auld Lang Syne" – before the book fell open at page 120. She read slowly:

Thou hast left me ever, Jamie!
Thou hast left me ever;
Thou hast left me ever, Jamie!
Thou hast left me ever.
After hast thou vowed that death
Only should us sever;
Now thou's left thy lass for ay –
I maun see thee never, Jamie,
I'll see thee never!

Thou hast me forsaken, Jamie!
Thou hast me forsaken;
Thou hast me forsaken, Jamie!
Thou hast me forsaken.
Thou canst love anither jo,
Whilst my heart is breaking;
Soon my weary een I'll close,
Never mair to waken, Jamie,
Never mair to waken!

She closed the book and brought it up to her bedside locker.

130

CHAPTER EIGHTEEN

BRIDIE KEPT TO THE HOUSE AS MUCH AS SHE COULD, BUT OCCASIONALLY she had to serve in the shop during the daytime when Podgy was away and Hanratty was not available. She was nervous and unsure at first but when she grew accustomed to the task she actually enjoyed it. It was a novel and welcome change from the drudgery of cleaning the house. The customers' reactions varied. The men in the bar enjoyed the respite from Podgy's gruff manner. The women who came to the grocery treated her with varying degrees of caution and distance. Their manner of address ranged from the polite "Miss McKevitt" to the familiar "Bridie." Lizzy Rogan made no effort to conceal her feelings about Bridie's presence. She went "Tsk! Tsk!" as she worked her way through her shopping list and on leaving the shop she shook her head and voiced her disapproval loudly to no one in particular.

"Disgraceful! An absolute disgrace!"

One caller whom Bridie was glad to see was Minnie McCabe, Jimmy's mother. Although her manner and tone put Minnie very definitely in the disapproving camp, she shared with Bridie a growing concern for Jimmy.

"Not a word from him this six months. And you?"

"Nothing," Bridie said.

"He has me heart scalded not knowing where he is or what he's up to. I say a rosary for him every day. Just for him."

"He'll turn up one of the days, you'll see." Bridie felt the lack of assurance in her own voice.

"I don't know. I was never in favour of this America business anyway. A big Godless country. Why he had to go there I'll never know." She packed her groceries into a tattered shopping bag. "It's not as if we drove him out. He should have settled down with a sensible woman for a wife," she added.

"How are the rest of the family?" Bridie asked.

"Och, they're all right. I hear your Pete is poorly."

"He is."

"I'm right sorry to hear that. He was a grand lad. And all he done for the football. I'll pray for him too. Good day now!" She moved towards the door.

"Good day Min – Mrs McCabe." Better to remain formal, Bridie thought.

"Mourning and weeping in this vale of tears," Minnie muttered as she edged sideways through the door.

Bridie visited Pete daily. It took her a few weeks to put her stamp on Podgy's house so, when Brian started attending school, there were only the routine household duties to be done each day. At ten o'clock each morning, she began the slow uphill journey to her home. On steep stretches of the journey she dismounted and wheeled her bike along the rutted roadway. Above her the beech and hazel leaves had turned russet and yellow and, on a particularly windy morning, they showered all about her on McMahon's hill. The hedgerows were draped with blackberry-laden brambles. Bridie spent an hour one morning filling two large cans with blackberries. She would bake a few blackberry and apple tarts for Pete. He always loved them. She remembered scolding him last autumn for eating almost a whole tart behind her back when he came in after a long day spent picking potatoes. Was it only last autumn? How strong and well Pete was then. And now?

"Mother of Jesus!" The shriek came from the other side of the hedge, where Maggie McMahon had been relieving herself. "Oh, thank the sweet God 'tis only yourself, Bridie McKevitt." She rearranged her clothing. Bridie turned away embarrassed.

"I'm sorry," she mumbled. "I was just picking some blackberries for – "

"For Pete, is it? How is poor Pete?"

"Not the best, Maggie."

"Ach the craythur. I was just thinkin' of him last night. The boys came home all down in themselves. Weren't they beaten in the first round of the Feis Cup! And them after winnin' it last year. I ses to them, I ses – yis can do nothin' without Pete McKevitt! And they agreed with me!"

Bridie hung a can of berries on either side of the handlebars and prepared to move off.

"D'ye know," Maggie said. "I have just the thing'll make him sit up. Hold on there and I'll bring it out to you!" Before Bridie could reply, Maggie took several large purposeful strides towards the house. She emerged a couple of minutes later with a bottle in her hand.

"Goat's milk! The doctors swear by it for any kind of weakness in the chest!" She thrust the bottle into Bridie's hand. "Tell Pete that'll have him kicking a ball again in no time! Good luck now!" She turned and again her great strides defeated Bridie's intentions.

Bridie coaxed her brother into sipping some of the goat's milk.

"Holy God!" He winced at the taste. "That's as strong as Maggie herself! I mind the time we used to tease that oul' goat of hers on the way home from school!" He managed a weak smile which cheered Bridie.

"It's hardly the same goat!"

"Knowing Maggie, it wouldn't surprise me," said Pete. "And goin' by the taste of this stuff I'd be damn near certain of it!"

It was a long time since Bridie had laughed heartily in Pete's presence, but she did so now.

"I had a visitor already this morning. You just missed him."

"Who?"

"The Canon. He was doing his rounds, visiting the sick." A further sip of the milk made him splutter, then cough. He pushed the bottle away from him. "Ah Jesus, that stuff'd kill you before it'd cure you!"

"Did he stay long?"

"The Canon? No. Just gave me communion and a blessing. He was a bit ill at ease. Kept looking off in the distance. I didn't detain him!"

Bridie picked some apples from an early windfall.

"I'm going up to make you a blackberry and apple tart," she said. "And you're to eat all of it!"

"At least it'll take the taste of McMahon's goat out of my mouth!"

133

Josie was singing at the top of her voice as Bridie entered the kitchen.

"Well indeed, someone is in good form here!" Bridie announced as she clattered the cans on the table.

"God, Bridie! You frightened the life out of me!" Josie said, clasping her hand to her breast. "I like to sing while I'm working!"

Bridie could not remember her singing like that when she was in the house.

"I promised Pete I'd make him a blackberry and apple tart. It's his favourite."

"Of course. Do you want a hand?"

"Not at all. You go on with your work."

"I'll make a drop of tea."

Bridie opened the press where she kept the baking dish. It was no longer there. A search of other presses revealed that everything had been changed around. Where once she could put her hand on the baking requirements without looking, a prolonged search was needed now. Josie sensed her frustration.

"Can you find everything all right? I did a bit of a clean-up."

"I'm managing rightly," Bridie said. She was a stranger in her own kitchen.

Later, while the tart was baking, she sat by the fire to take tea with Josie. She watched Josie's face intently.

"You had a visitor, I believe?"

"Och, don't remind me. I hadn't a dish nor nothing washed. The kitchen was a mess. No wonder he didn't stay long!" She sipped her tea. "He was asking after you."

"I dare say."

"Mmm. Said he must call down to see you." She propped a sod of turf on either side of the fire.

"You have something else to sing about, haven't you, Josie?" Bridie asked quietly.

"You noticed?"

"I could tell by the glow in your cheeks. When is it due?"

"March, please God."

A figure darkened the door of the kitchen. It was James.

"The smell of that tart is drivin' me mad up in the loft!" he said. "Is there any chance of a bit?"

"It's for your brother but if you have a few tales from the loft, I might just keep a slice for you," said Bridie.

A week later, Bridie was taking ashes to the dump in Podgy's garden when she heard a car approach. It was black and shining new. Fear gripped the pit of her stomach. Podgy had gone to Carrickmacross and she felt suddenly vulnerable. She was relieved to see Canon McKenna, and not Winston Pollock, emerge from the car.

"Well is this where you've been hiding yourself, Miss McKevitt?" The formality of the address boded ill. He tapped the bowl of his pipe on the gate-pier and patted his pockets in search of his tobacco pouch.

"This is where I've been, Canon," Bridie replied. "Will you come inside?"

"I think not. What I have to say can be said here." He sucked on the pipe before proceeding to fill it. "Now tell me. What are you about here?"

"I'm protecting my nephew, Canon."

He packed the tobacco into the pipebowl with two fingers.

"Noble intentions do not excuse questionable acts."

"I don't understand, Canon."

"Of course you understand, woman. You are young, single and unattached and you are living under the same roof with a single man."

Bridie was not prepared for this onslaught. She stared at the priest, as he reddened his pipe.

"I am his housekeeper, Canon."

"Housekeeper! Housekeeper! Is that what people will think, *do* think?"

"I can't help what people – "

"Well I can, and I will not have you giving scandal to the parish."

"There is no – "

"Not to mention the scandal given to a young boy of tender years. 'Woe to the scandal giver,' the Lord said – 'it were better that a millstone be tied around his neck' – "

"What I'm doing is for the boy, Canon."

"What you're doing is wrong, woman!" He pointed the stem of his pipe menacingly at Bridie. His hand trembled slightly. "I repeat.

135

Noble intentions do not excuse wrong actions! You must extricate yourself from this situation, woman."

Bridie looked at the ground.

"I am not going back home, Canon," she said in a whisper.

"Speak up, woman!"

"I said I'm not – "

"Well then you must find an alternative lodging – where there will be no scandal. There are widow women in this parish who would be glad of company." He fumbled for his matches. "Or you could legalise this situation," he continued. "Peter Kernan would make an excellent husband – " He lit his pipe again " – and stepfather, I am sure. Either way, this situation must not continue." He paused, pointedly. "I hope you will be more heedful of my words than your sister was." His words stung Bridie. She could not stop her tears.

"I appreciate your motives for the boy but he is the very one you are damaging. I expect the situation" – again the pause – "to be rectified forthwith. Otherwise you will be read off the altar. Good day."

Pete McKevitt died on October 24th 1930. In his last weeks Hughie carried him into the house each night to sleep. "It's a poor day if we bring the cows in at night and we don't bring in the humans," he said.

Pete's funeral was the biggest Cullyboe had seen for many years.

"The Canon'll be happy tonight," Podgy said afterwards. "Such a rake of offerings he took in. The new car is well paid for now, I'd say."

"Are you expecting that white house back?" Hughie asked him.

"Sure what would I want with it?"

"That's all I want to hear," Hughie said, "because there's nothing I detest more on this earth than that . . . prison!"

Three nights later Hughie poured paraffin oil all over the white house and set it alight. The blaze could be seen all over Drominarney. As the neighbours came to their doors and watched, Josie McKevitt stood in her doorway and played "Teddy O'Neill" for the very last time.

Five months later Josie gave birth to a boy. He was christened Peter Joseph McKevitt.

CHAPTER NINETEEN

BRIDIE LEANED HER CHEEK AGAINST THE WARM BELLY OF THE COW AND closed her eyes as she listened to the even rhythm of the milk singing into the bucket she held between her knees. There was great comfort in the animal's warmth. Bridie spoke soothingly to the cow as she had seen her father do many years before – a strange lilting tune that made no sense to the human ear but which united man and beast in a perfect way . . . As if in response to Bridie's talk, the cow would occasionally flick her tail around to toss Bridie's hair about her head.

"Here she is folks – The Queen of the Cowshed!" The voice startled Bridie out of her reverie.

"Jesus Christ!" she whispered.

"Well not quite, but the next best thing, I'd say!"

She jumped off the stool, all but spilling the milk from the bucket.

"Jimmy McCabe! Where in the name of God did you come out of?"

He stood there, beaming at her, his hands thrust into the pockets of a bright blue suit. He obviously enjoyed the effect of his surprise arrival. Bridie patted the rump of the cow which had grown restless at the interruption of her milking.

"In answer to your question, I jumped off the Belfast train at Culloville and came straight here to see you – never even called to the home place! Josie told me you were down here so – here I am!"

Bridie rested the bucket on a shelf and wiped her hands on her apron.

"Well, you're the cool customer!" she said after a long pause.

"None cooler, baby!" Jimmy laughed.

"You haven't changed a bit either, apart from the Yankee accent and the flashy suit!"

"You haven't changed yourself, Bridie. Still as hard as ever on me. It's just as well I have a thick skin – "

"Hard as ever? Hard as ever! You disappear for more than two years. Not a word or a sign from you until you turn up out of the blue, dressed like one of them film stars."

"Ah, so you've seen my movies!" His attempt at humour deepened the furrow on Bridie's brow.

He changed his tone. "Well, at least I did turn up!" he said meekly.

"After two years. Your poor mother was out of her mind, not knowing whether you were alive or dead – "

"And what about you? Were you – ?"

"Where were you? Why did you never reply to our letters? What were you doing?"

Jimmy was taken aback at the vehemence of Bridie's questioning.

"Easy on! Easy on! All in good time! It's a long story. But what are we doing talking here in the cowshed?"

"There's many a time it was good enough for you!"

"God above but you're the hard woman, Bridie McKevitt! Why are you doing the milking anyway? Where are the men?"

"Hughie's away at the fair of Carrick. Josie has a baby to mind and another one on the way."

"And Pete? How is the football star?"

Bridie took a big breath.

"Pete's in Cullyboe cemetery this two years."

Bridie regretted giving him the news so baldly. The colour drained from his face as he looked anxiously around for something to lean on.

"Christ!" he whispered. "Christ Almighty! I never knew. Pete. Pete dead! I'm terrible sorry, Bridie. Terrible sorry." He sniffed repeatedly as he fought back the tears. He turned quickly away and looked across the haggard.

"How?"

"TB."

"Jesus! Is there no end to the scourging of your family?"

Bridie did not answer but took down the bucket from the shelf.

Jimmy was surprised at her reaction. "I suppose the milking has to be done, whatever."

"Aye."

"Look, Bridie, I'm terrible sorry about – about everything. Hearing about Pete is just too much. I'll come round this evening and – "

"I'll not be here."

"Well, where will you be, then?"

"I'll be at Podgy's."

"At Podgy's? What in God's name would you be – "

"I live there."

"You live there?"

"Yes. Podgy and me were married two years ago."

Jimmy stared at her. He groped like a blind man for the lintel of the doorway for support. He took a deep breath. "Married?"

"Yes."

"To Podgy Kernan?"

"Yes."

The milk sang a softer tune as it squirted into the rising froth.

"What did you want to do that for, in the name of God?"

Bridie stopped milking momentarily, then resumed with greater urgency.

"What does anyone get married for?"

"Ah for Christ's sake, Bridie! He's at least twenty years older than you. He lives in a pigsty of a place. He's a mean oul' chancer that wouldn't give you – "

She shoved the stool back noisily, slammed the bucket down on the floor and turned to face him.

"He gave me – and Brian – a roof over our heads when we needed it. He did many's the kind deed for all of us over the past years when we had our troubles. The pigsty – as you call it – only needed a woman to clean it up. Life wasn't easy for him either, you know. So what if he's older than me? He was a good friend " – she paused and looked defiantly at Jimmy – "and he was here!"

The last remark stung him. He turned away and drew a kick at the door. "And I was on my holidays in America!"

"If you'd read my letters you would have known – about Pete, about a new bride in the house, about the Johnstons –"

"What about the Johnstons?"

139

"They tried to take Brian back!"

He turned to her again. "You did it for him, didn't you? For Brian."

She picked up the bucket.

"I had decisions to make about my own life – too." She moved towards the door.

He went out into the haggard ahead of her. "Right! Well I hope you'll all be very happy! I'll see you!"

He strode across the yard and onto the lane where a pony and trap were waiting.

Bridie stood for a long time listening to the echo of the pony's hooves receding in the still evening.

She lay in bed that night thinking about the shock and the pain of Jimmy's visit. Yes, she had meant what she had said. Maybe she had been blunt but he'd been out of contract for two years. What did he expect? Two years. She could hear Podgy shuffling about downstairs. He lived there now, hobbling from the little bedroom off the kitchen to his chair by the range where he spent most of the day. He rarely visited the bar and shop now. When he did, it was only to pick an argument with a customer or to insult everyone present. Bridie would shepherd him back to the kitchen and return to serve in the bar.

Podgy's health and general mood had deteriorated over the two years. His swollen and discoloured foot caused him constant pain but he stubbornly refused to have it examined.

Two years. Two years of marriage. Marriage? Or "legalising the situation." Wasn't that what the Canon said?

Marriage? Three days in Dundalk. Podgy wouldn't trust Hanratty in charge of the shop for any longer. Days, chilly days, spent at Blackrock sitting watching people go by, children paddling, sheltering in a tea-room. Nights of groping, roughness, fumbling, slobbering, failure. Nights of waking to deep snoring, easing away from a sweating shirt-clad body, gripping the edge of the bed and gazing through threadbare curtains at the moon. Remembering when she and Sarah, as children, would sit up in bed on moonlit nights and make up silly rhymes.

"Man in the Moon, Man in the Moon
Why did the dish run away with the spoon?"

"Hey diddle diddle the cat did a piddle
On top of the Man in the Moon."

Their uncontrolled giggling would eventually land them in trouble when their mother would bang on the door and shout, "If youse don't stop that I'll send for the bogeyman! . . . "

Another snort from beside her, a wild looping swing of an arm and the bedclothes were dragged away from her. She forced a smile. The bogeyman had come at last . . .

Within a month he had moved out of the bedroom. "Can't manage that stairs with this curse-of-God foot . . . " He became increasingly withdrawn and cranky and made no effort to run the shop. Bridie took over the business but she still had to hand over the takings in the biscuit box. Not before she managed to leave a little aside each night that would add up to a little extra a week for Hanratty. He was a simple lad, but an honest worker if handled gently, something which Podgy could never achieve. At least Brian had settled in well at school. She kept him out of Podgy's way and during the school holidays the boy loved to help Hughie on the farm or play in James's workroom.

And now, out of the blue, Jimmy McCabe had come back. Another bogeyman. She reached across for *The Works of Robert Burns*. She knew that if she opened it, it would fall open at a well-thumbed page. She put the book back and turned over with a sigh.

"Man in the Moon, Man in the Moon
Why did the dish rush run away with the spoon?"

A week later a letter arrived. Bridie recognised the handwriting even before she noticed the telltale crossing out of "McKevitt" on the envelope, replaced in a scrawl by "Kernan". There was no address on the letter.

Dear Bridie,

Just a short note before I disappear again. To say I'm sorry for leaving you in a huff like that. I suppose we gave each other a shock – me turning up like that and you announcing you were married. I couldn't get over that. Still can't. I suppose you have heard by now that I nearly drank Martin's pub dry in Cullyboe. I don't think my poor mother saw me at my best when I eventually did get home.

Anyway it's over now. Drink never solved anything, but as a mate of mine once said, "It sure is nice to try!" I don't know where I'll go from here. East or West? I don't know whether I can face the mines again. That's where I ended up. In the Appalachian mountains. The wildest place God made. I worked like a black there for the two years. Never got out of it once. I gave a letter to a fellow one time but he obviously never posted it. And of course I never got yours either. I left no forwarding address, as they say over there.

Anyway that was the way of it. All I can do is wish you luck. From the sound of things you'll need it. Say hello to the boys for me.

Regards,

Jimmy.

PS The one place I did go when I was sober was to Pete's grave. Christ, even though I read it on the headstone, I still can't believe it. God rest him.

PPS Sorry for the mess on the envelope (only one I can find!). To me you'll always be Bridie McKevitt.

J.

CHAPTER TWENTY

BRIDIE HAD JUST REACHED THE SCHOOL GATE WHEN THE CHILDREN CAME charging out the door. She recognised instantly Brian's fair head bobbing among the excited children.

"Aunt Bridie! Aunt Bridie!" The boy wove his way through the throng. "I'm going to Dublin! I'm going to Dublin!"

"My goodness! Are you going to walk all the way?"

"No!" He tugged at her skirt in disgust. "Master Lennon is bringing three of us to the Euch-ar-istic Congress." He had obviously rehearsed the strange word. "Me and Micky McMahon an' Maureen Brady. 'Cos we knew our catechism for the priest and all about St Patrick. The master said he'd bring the three best on the train to Dublin!" He sensed the doubt in Bridie's expression. "If it's all right with you," he added warily.

"Well, we'd better find out a bit more about this," she announced, taking the child's hand as she marched back into the school.

"They'll be perfectly safe, Mrs Kernan. Mrs Lennon will be coming with us, so we thought we'd give three of the first communicants a treat. It will be a great experience for them. Truly it will."

Master Lennon was an earnest man in his forties. He and his wife were the only teachers in Drominarney School and were themselves childless. Both were extremely pious. They cycled to seven-thirty Mass in Cullyboe every morning. Séamus Lennon was at the core of every cultural and religious movement in the parish – the sodality, the Pioneers, the Gaelic Athletic Association.

143

For all that, Bridie was uneasy about the trip to Dublin but in the end she relented. She dressed Brian in his smartest clothes and, when the Lennons came to collect him, she stood watching and waving as the pony and trap, driven by Charlie McMahon, disappeared in a cloud of dust.

The children found it hard to contain their excitement, straining eyes and ears to see who would be first to spot the Dublin train.

"Save your energy, children," Master Lennon said, as he consulted his pocket-watch. "We're way early. The train'll not be here for another half-hour."

The woman on the opposite platform watched the children with growing curiosity. She summoned a young fair-haired girl to her side, took her hand and proceeded to the pedestrian bridge. She hesitated and glanced across at the children once more before ascending the bridge. When she descended on the opposite side she was sure.

"It's Brian, isn't it?" she called.

Brian was playing a game of tig with his friends, but he stopped suddenly when he heard his name.

"Brian, Brian Johnston!" The woman called as she approached him. Master Lennon sensed danger and broke off his conversation with his wife to confront the strange woman.

"Is there something the matter?" he asked cautiously.

"I just recognised Brian from the other side," the woman said. "I'm his Aunt Rose." She noticed the muscles tensing on the schoolmaster's face.

"It's all right," she assured him. "He's in no danger. Hello, Brian!" She turned to the boy who gave her a puzzled look. "I don't suppose you remember your Aunt Rose! And all the times I minded you as a baby!"

The boy squinted in the sunlight and shook his head.

"Well I am your Aunt Rose, your Daddy's sister. And this – " she tugged the hand of the girl who had been hiding shyly behind her " – is *your* sister. Your twin sister, Hannah."

The children stood facing each other, silent, shy. Brian felt a hand cup his elbow.

"Are you not going to shake your sister's hand?" Master Lennon whispered, as he pushed Brian's arm forward.

They shook hands awkwardly. The adults withdrew as Master Lennon introduced his wife to Rose.

"We're going to Belfast, to my grand-aunts, for two weeks," Hannah said suddenly. "Where are youse going?"

"To Dublin. To the Euch-ar-istic Congress." He measured the word carefully.

"What's a Euch – ?" She shrugged her shoulders.

"It's to celebrate St Patrick bringing the faith to Ireland fifteen hundred years ago!" Brian recited the answer he had learned for Canon McKenna's visit. "I knew that for the priest – "

"Taigs. They're all Taigs in Dublin. That's what Winston says." Hannah's tone changed from curiosity to disdain.

"What's Taigs? Who's Winston?"

"Taigs is Catholics – like you." Her voice was imperious. "And Winston's my uncle. He's married to Aunt Rose."

"Where do you live?"

"In Sloe Hill. Where do you live?"

"With my Aunt Bridie, in Drominarney. Can I come to see you?"

"'Spose so. I'll ask Aunt Rose."

The whistle of an approaching train caused Rose to turn anxiously, take Hannah's hand and hurry her back across the bridge.

The Belfast train arrived in a swirl of smoke and steam. Brian watched and waited until he saw Rose and Hannah move through the train and take a seat at the rear of the second carriage. Hannah smiled and waved to him. He waved back as the train began to move forward slowly, steam hissing from under the engine.

"S-s-ister, s-s-ister, s-s-ister," it said.

It was Brian's first journey on a train. He sat by the window mesmerised by the patchwork fields which slid by, until Maureen Brady gave a shriek from the other side of the carriage.

"The sea! The sea! Look at the sea!" They gazed in awe at the vast expanse of water. None of the children had seen the sea before.

"I wouldn't go out in that if you paid me," Micky McMahon said.

At Dundalk and Drogheda the platforms were thronged. The heat in the crowded train was overpowering. Their discomfort made the children restive. Even the train's slow progress through the suburbs of north Dublin failed to distract them. Master Lennon heaved a sigh of relief when the train finally rolled into Amiens Street station.

Dublin on a bright warm June morning was a wonderland of colour. Talbot Street was festooned with bunting. Papal flags hung

145

from every building and shop-windows had been cleared of their displays to make way for pictures of the Sacred Heart, statues of St Patrick and a variety of Eucharistic Congress banners. The children, holding hands under strict instructions from Master Lennon, turned this way and that as each in turn discovered a new source of surprise.

"Look at this!"

"Look at yon!"

They pressed their faces against the window of a butcher's shop which featured a reproduction of the Lourdes grotto, complete with a stream running through.

"Our Lady of Lourdes, Pray For Ireland," the legend said.

"Come on, children." Master Lennon beckoned. "We have to get a good position for the ceremonies."

A huge altar had been constructed on O'Connell Bridge. A vast crowd had assembled before it. Master Lennon led his charges down an alleyway which opened on to the quays. They wormed their way through to the riverside wall from where they had a reasonable, if distant, view of the altar.

A great choral swell announced the beginning of the ceremonies as a procession of bishops and priests surrounded the altar. Brian was startled by the voice of the celebrant which boomed through a loudspeaker affixed to the lamp post beside him.

An elderly woman closed her eyes, her lips moved in rapid prayer as she fingered her beads ceaselessly. Beneath him, boats rode at anchor on the river, their occupants straining to get a clear view of the altar. The ceremonies wore on. When the sermon began, the woman beside Brian continued to pray her rosary, eyes closed, her gnarled fingers passing over the beads at an even faster rate. Brian turned his attention to the seagulls riding the waves of the river. He focused on one particular bird and followed its movements with interest. It was gradually swept back by the tide until it rose and flew forward to where it had been a few minutes previously.

A firm tap on his shoulder woke Brian from his reverie. Master Lennon motioned towards the altar and loudly encouraged Brian to respond to the litany. The old woman continued to finger her beads.

When the ceremonies finally ended, the crowd dispersed in an atmosphere of great joy. For the three children, this was the most exciting part of the day. They moved among the vast, noisy, gay

146

crowd, caught up in the pomp and pageantry of the occasion. Master and Mrs Lennon took them down Westmoreland Street and pointed out various landmarks – the Bank of Ireland ("once the House of Parliament"), Trinity College ("the Protestant university"). They made slow progress up Grafton Street, pausing to peer in at the window displays in the fashionable shops. Finally they reached St Stephen's Green where Mrs Lennon conjured a magical picnic from the basket she carried with her. They fed the scraps to the ducks in the pond nearby.

"Them's not ducks at all," Micky McMahon announced. "My mammy has ducks at home – big white ones. Them's not ducks at all!"

Master Lennon treated the children to ice-cream and lemonade in the Monument Café. They were by now sated with the delights of a long day. Before they trudged up the steps of Amiens Street station, Master Lennon made one more stop – at a kerbside stall just outside the station. As they waited to board the train, he presented each child with a memento of their day out – a gleaming "Eucharistic" medal.

"This is a very special day in your lives, children. You will remember it for the rest of your lives. I want you to have this souvenir of the Eucharistic Congress to remind you of this day and of the love of Our Lord Jesus Christ in the Blessed Eucharist – the Eucharist you will receive for the first time in a few weeks. God bless you, children."

"Thank you, Master," the three chorused spontaneously.

"Mind the children, for God's sake," Master Lennon appealed as the crowd surged aboard the train. "No manners," he muttered as he shepherded his charges into the carriage. Brian hoped to find a window seat but, in the end, he was lucky to share a passageway seat with Micky McMahon. He craned his neck to get a last glimpse of Dublin. A burly man by the window opened out a newspaper and shut out the view. Brian leaned back in the seat. He would have so much to tell Aunt Bridie about all he had seen. The guard's whistle sounded. There was a toot from the engine. Steam swirled from beneath the carriage.

S-s-sister! S-s-sister! S-s-sister!

"Race you to the bottom of the brae!" Hannah called.

*"That's not fair. You have a start," Brian cried, but Hannah was
already careering down the newly-cut meadow, her blonde ringlets
dancing in the breeze. "Not fair," he sobbed but Hannah only taunted
him further.*

*"Slow coach! Slow coach!" A man and a woman waited at the
bottom of the field, their arms outstretched to greet the children. They
were laughing but made no sound. Brian could not distinguish their
faces as he fought in vain to overtake his sister.*

S-s-sister! S-s-sister! S-s-sister!

*Hannah bounded into the man's arms. He swung her around in
triumph. Brian collapsed in tears into the woman's arms. She pressed
him close to her. She felt warm and soft. Her hands cupped his face.
Gentle hands. He looked directly at her, but her face seemed to be
surrounded in a haze. Hannah chuckled helplessly as her father
tickled her. Why could he not hear the man or the woman? He knew
they were speaking, yet he could hear nothing. The woman lifted him
up. He buried his head in her flowing hair. It smelled of flowers. Her
hair dried his tears. At last he heard something other than Hannah's
laughter. A toot from the horn of a car. A black car had drawn up on
the road outside.*

*"Hello, Uncle Winston!" Hannah called. A big man got out of the
car.*

*"Hello, missy!" his voice boomed. Brian felt suddenly cold. It was
the bad man. Why could he hear the bad man and not . . . ?*

*They moved towards the car. Brian trembled but the woman's firm
grip reassured him. They climbed into the back seat. He sat on the
woman's lap. Hannah made a face at him from the man's lap. He
made a hissing noise back at her.*

S-s-sister! S-s-sister! S-s-sister!

*The car stuttered into life as the bad man swung the starting handle.
He clambered into the driving seat. He seemed to occupy the entire
front of the car, almost blocking out the light. The car moved forward.
The woman dandled Brian on her knee to the uneven movement of the
car. He felt better.*

Bumpedy-bumpedy-bumpedy-BANG! Suddenly the windscreen shattered and glass showered all over the front of the car.

"Bloody Taigs!" the bad man screamed. The car jerked to a halt as he clutched his face in obvious pain.

"Bloody Taigs!" he roared again as he turned around. Blood streamed from a huge gash on his forehead and seeped out through his enormous fingers . . .

Brian could hear screams all about him. He was staring into the bloodied face of the big man, not the bad man, but the big man who sat by the train window. Outside he could hear angry voices.

"Bloody Taigs! Go back to Dublin the lot of yis!"

"Aye and kiss the Pope's arse!"

More breaking glass. The train was moving very slowly through a cutting. Suddenly Brian felt himself being wrestled out of the seat and onto the ground. Master Lennon lay on top of him.

"Stay down!" he screamed above the din. "We're just coming towards Culloville. We'll be all right in a minute or two." Almost as he spoke, the train picked up speed and was soon clear of the cutting. The noise gradually subsided as people picked themselves up and carefully removed shards of glass from their clothing. Handkerchiefs were proffered to the big man.

"I'm a nurse! I'm a nurse!" a voice called. A path was cleared to the big man who pressed several handkerchiefs to his head in an effort to staunch the flow of blood.

"Young pups! Bowsies!" Master Lennon muttered. Brian stood beside the master. The nightmare was over.

For over an hour, Bridie watched anxiously for Brian's return. As soon as she recognised Charlie McMahon's trap approaching, she dashed out onto the road.

"Yis are very late!" she called.

"The train was very crowded and we had a bit of trouble coming towards Culloville. Young Proddie pups broke a few windows," Master Lennon replied. "But we're all right and we had a great day, hadn't we, children?" The three children nodded and yawned as one.

"Ye'll all sleep well tonight," Bridie said as Brian fell into her arms. "Thanks, Master."

The trap moved off. "Oh – and Brian has a big surprise to tell you about!" Master Lennon called as the trap was enveloped in a cloud of dust.

Bridie took Brian's hand.

"What's this big surprise?" she asked.

"I met my sister!" Brian stifled another yawn.

"Where?" Her grip tightened on the boy's hand.

"At the railway station. She was going to Belfast – with Aunt Rose."

"What did you do?"

"We just talked. Can I see her again?" He gave a long yawn.

"We'll see. Right now, it's time for your bed. You can tell me all about your day out tomorrow."

As they turned to go into the house, a gruff voice came from the kitchen.

"Is anyone gettin' supper in this house tonight?"

PART TWO

CHAPTER TWENTY-ONE

MOIRA OPENED THE DOOR GENTLY. SHE PAUSED TO SEE IF THERE WAS ANY reaction from the figure in the bed.

"Is that you, Aunt Rose?" The tiny voice was muffled by the sheet she had drawn around her.

"No, Mum. It's me. Moira. I have a visitor for you. It's Uncle Brian." She motioned Brian forward to the bedside.

"Hello, Hannah," he said softly. He laid a bouquet of flowers on the folded coverlet.

Hannah focused on the figure that seated itself at the bedside. It was a slow and painful process, interrupted by thunderous waves which crashed through her brain with fearsome regularity. She could see each wave approach and rise into ogreish proportions before pounding itself to destruction against the eggshell wall of her skull . . .

She shut her eyes tightly, then opened them slowly to welcome the brief respite that would follow. The rotund figure of her brother came slowly into focus. A hand came from behind her, gently combing her hair with its fingers.

"I'll leave you two together," Moira said. "And steep these lovely flowers. Did you see them, Mum?" She came around to hold them in her mother's line of vision.

"No need. No need for flowers."

Moira withdrew and closed the door.

"I'm sorry I'm so late," Brian began. "There was a bomb – near Victoria Station. Everything was – "

JOHN QUINN

"Your crowd again . . . "

"They're not my crowd, Hannah. They never were. You know that – "

Wrong. Wrong. He should never have mentioned the bomb. He should know by now. Begin again.

"Well, how are you feeling? You look a little – "

"Worse. I'm dying. Slowly."

"Are you in pain?"

"Am I in pain? Am I in – " The wave rose up, assumed another fiendish shape and hovered, before dashing itself into the wall, with even greater intensity . . .

"A drink?" Brian whispered.

She shook her head imperceptibly.

He sat silently for a minute or two, allowing her time to recover. He looked at her lined face, the slight figure which hardly disturbed the bedclothes. It seemed as if she had shrunk to a third of her size.

He reached into the pocket of his jacket.

"I'll pray a little with you," he said.

"No need. No need for prayers." She took a long slow breath. "My suffering is my prayer. Isn't that right, Aunt Rose?"

He opened the book and flicked through the pages. He paused at Psalm 102.

O Lord, listen to my prayer
And let my cry for help reach you.
Do not hide your face from me
In the day of my distress
Turn your ear towards me
And answer me quickly when I call.

His eye wandered to the top of the bedside locker. A little brass bell. A glass of water. An old sepia-toned photograph in a frame showing a young man and woman posing shyly on either side of a haycock. He recognised them with some difficulty. Gordon and Rose Johnston.

For my days are vanishing like smoke.
My bones burn away like a fire.
My heart is withered like the grass.
I forget to eat my bread.
I cry with all my strength
And my skin clings to my bones.

154

A pretty eggcup with a rose design, decorated with a strange oriental pattern.

* * *

Eggs. Eggs smashed all over the kitchen floor . . .

"Hannah, child, what did you do that for?" Rose's voice expressed both anger and puzzlement.

Hannah looked with some delight at the eggwhites merging into each other on the flagged floor.

"Uncle Winston says we don't need eggs from Taigs."

Rose turned away in embarrassment.

"Och, Hannah!" She grabbed a brush from behind the door and began sweeping the mess towards the door.

"I'm – sorry about this," she said.

"It's all right." Bridie's voice was hoarse. "At least the dog will have a good feed." Rose opened the back door and swept the mess over the threshold and onto the street where a lean emaciated sheepdog approached it cautiously and then began to lap it up.

Bridie had deliberated for a long time about this visit. Ever since Brian had met his sister at Culloville railway station he had pestered Bridie about seeing Hannah again. He spoke of having funny dreams in which Hannah, and strange people he did not recognise, appeared. The boy's persistent questions did not endear him to Podgy.

"Does he ever talk about anything only his curse-of-God sister?" he muttered from his chair in the corner.

"It's only natural he'd want to see her again," Bridie reasoned.

"I'm tellin' ye. There'll only be trouble if ye go up to that place. Go easy, will ye, for Christ's sake."

Twice a week Bridie had the distasteful task of changing the bandage on Podgy's leg and bathing its weeping pus-filled sores. Her hand trembled as she dabbed the sores with a pad of lint. Podgy gripped the arms of his chair.

"Jesus Christ Almighty," he muttered. "Will ye go easy, I said."

"I can't do any better. You'll have to go into hospital to get it looked after properly like Dr McMahon says."

"Dr McMahon, Dr McMahon. I wouldn't go next or near them cronies of his – all out to make money."

155

Brian burst excitedly into the kitchen.

"Aunt Bridie, could Hannah come to my First Communion?"

"Ah, for the love of Christ!" Podgy roared. "Am I to be scalded with this – day in, day out?"

Bridie became increasingly troubled by Podgy's perverse attitude to the boy. She called on James in his saddlery.

"I know he's in a lot of pain but it's not good for Brian to be in the middle of all that," Bridie said.

"Would you not think of moving him back here?"

"No. Josie has her hands full."

James drew on a cigarette and exhaled slowly.

"There's Drumcran," he said.

"The sisters?"

"Aye."

Drumcran was an orphanage and school run by nuns in a converted estate house about twelve miles south of Cullyboe.

"I – hadn't thought about that."

James could sense that his sister was taken aback by his suggestion.

"It would only be for a while – till Podgy gets better."

"I don't know anyone there," Bridie said.

"Petesy Marry's cousin is the Reverend Mother there. I could find out – "

"Would you, please? There's something else . . . " She told him of Brian's meeting with Hannah and of the boy's repeated requests to see his sister again.

"What do you think?" James took another cigarette from the packet.

"He has a right, I suppose. It's playing on his mind a lot."

"Do you think they've cooled down?"

"Well, there's been no contact for a long time." She wound a length of waxen thread around her finger and unwound it again. "And Master Lennon said that Rose was grand."

"Well maybe a visit would be all right. Do you want me to go with you?"

"No. I'll be fine."

"More tea? Would you like more tea?" Rose's voice woke Bridie from her reverie.

"I'm – sorry. No thanks."

Brian tugged at her skirt.

"Aunt Bridie. Come on and see the wee calf. He was only born yesterday."

"All right. But we have to go home then. Podgy'll be wanting his tea."

She followed the twins across the haggard to a long shed. The children were in animated conversation.

"He's my calf. Uncle Winston's going to rear him specially for me," Hannah boasted.

"I have my own wee hammer in Uncle James's loft," Brian said. "He lets me play with it every time."

"I'm going to feed the calf every day."

They made their way along a dark passageway through the shed until they came to a small makeshift pen.

"There he is, Aunt Bridie. Isn't he a brave wee fellow?"

The white-faced calf struggled to its feet on their approach. The children petted it and shrieked with excitement as the calf tried to suckle their fingers. Its mother in the adjoining pen lowed quietly. Bridie joined in the children's fun. Unnoticed by the trio, a burly figure emerged from the gloom at the far end of the shed.

"Well, young Johnston, have you decided to come home at last?"

Winston Pollock's voice sent a chill through Bridie. She clasped Brian's arm.

"We're just on a visit," she said firmly. "We were about to go home."

"This will always be the boy's home. You denied him that."

Bridie ushered the children out of the shed and marched Brian across the haggard. She called out a farewell to Rose and kept going towards the pony and trap. Brian turned to wave to his sister. Hannah stood in front of Winston Pollock, his hands resting on her shoulders. Bridie felt his dark stare burn into her as she took the reins and turned the trap. The pony galloped down Caraher's Brae, but Bridie didn't dare look back. Drumcran was more than a mere possibility for Brian, she thought.

JOHN QUINN

Canon McKenna turned to address the children who sat in the front pews of the church.

"Today is a very important day in your lives, children. You are about to receive the body and blood of Our Lord and Saviour, Jesus Christ, for the very first time. This is a great privilege for you. If a great person like Mr de Valera or the Pope were coming to visit your house, there would be great preparations. Your father would be out whitewashing and your mother would be scrubbing the floors so everything would be spotless for the visitor. Well now, each of you will receive the greatest visitor of all today – Jesus Christ, the Son of God – and that is why you have made great preparations with your teachers and your parents and why your souls have been made spotless by the sacrament of confession.

"Holy Communion will be food for your souls throughout the journey of life, children. If you receive it as often as you can, it will keep you pure and good. It will help you resist the temptations of the world, the flesh and the devil. There is no better example for you than a man who is in our midst this morning, a man who was like you twenty-five years ago. Pat Trainor – I must correct myself – Father Pat Trainor is back in his native parish and will say his first Mass here at eight o'clock tomorrow morning. We all rejoice with him and wish him well when he leaves for Africa next week to begin the great work of converting the pagans of that dark continent."

Master Lennon had difficulty in gathering the children together for a group photograph in the churchyard. They chattered and ran about excitedly until they were finally assembled to face Mrs Lennon with her Brownie camera. Bridie looked with pride at the fair-haired boy with the serious expression. She waved and caught his attention. Just as the shutter clicked, he smiled nervously at Bridie. In that instant she could see her sister's impish face in his expression.

"Man in the Moon! Man in the Moon!
Why did the dish – "

"Isn't he a grand boy? The image of his mother!" Pat Trainor stood beside her. His black suit and collar looked a size too small for his stout frame and the smell of stale sweat almost overwhelmed Bridie.

"He is that, surely." She edged away from him as the group broke up.

Brian ran up to Bridie.

158

"Hello, Brian, and congratulations!" Pat Trainor offered his hand to the boy.

"This is Father Pat Trainor, Brian," Bridie explained. "He's just been made a priest." Brian shook hands with the priest. He had hands like Podgy. Wet hands. Brian instinctively drew his hand away and was amazed to find two gleaming half-crowns in his palm. He gazed at them, speechless.

"God bless you, Brian," Pat Trainor said as he slipped away into the crowd.

CHAPTER TWENTY-TWO

"AUNT ROSE! AUNT ROSE!"

Hannah's terrified screams brought Rose running to the child's bedroom.

"What is it, child?" Hannah sat upright in bed, her whole body shuddering. Rose swept her into her arms and drew a blanket across the child's shoulders.

"Did you have a bad dream?"

"Uncle Winston – was shouting – and shouting – about a child," Hannah sobbed. "Was he – shouting – about me?"

Rose cradled the child's head under her chin.

"Shush! You just dreamed it. There was no one shouting." She rocked the trembling body gently to and fro.

"I'm afraid, Aunt Rose."

"It's all right, child. I'm here now and I'll stay with you."

The door creaked behind them. Hannah's body went rigid. Through her tousled hair she could see the massive frame of Winston Pollock's body filling the doorway. He wore only a shirt.

"It's all right," Rose said quietly, without turning around. "Hannah had a bad dream. She's a wee bit upset. I'll stay the rest of the night with her."

Winston stood silently in the doorway for a long time before he turned away and slammed the door behind him.

Rose eased the child's head back onto the pillow and slid in beside her in the narrow bed.

"Now, isn't this nice and cosy?" she whispered. Hannah nodded.

"Is Winston cross with me?"

Rose ignored the question.

"Do you know what I forgot to tell you? We'll be going to Drumgore sports on Thursday and you might see Brian there! Wouldn't that be nice?"

Hannah nodded. As she did so her eyes closed slowly. Rose kissed her gently on the forehead. Sleep did not come easily for Rose. The narrow bed did not afford much comfort or ease of movement. Hannah's sleep became restless. She occasionally threshed out her legs in fitful movements. Rose grimaced in pain. She eased a pillow from under her head and slid it between her lower body and the child's legs. It would protect her bruised and tender thighs.

Drumgore sports were held in the local football field on the fifteenth of August each year. Despite the overtly Roman Catholic nature of the occasion – it was a church holy day, the feast of the Assumption of Mary – the sports had always attracted support from all sections of the community, primarily because the proceeds went to the upkeep of Drumgore House, a home for disabled veterans of the Great War. The creation of the border had diminished its appeal initially but, as the years went by, the support rose again, thanks mainly to the energetic teamwork of Father O'Hagan, the parish priest, and Mr Price, the rector.

James and Hughie McKevitt called for Bridie and Brian in the pony and trap.

"Let yis not be dallying too long at that oul' curse-of-God sports," Podgy called out. "There's work to be done here."

"No bother," Hughie shouted, giving a broad wink to Bridie. "We'll be home at midnight, Podgy!" He cracked the reins on the pony's back and the trap rumbled off before its occupants could hear the angry ranting from within the kitchen.

Drumgore was *en fête* for the sports. The tree-lined approach to the field was festooned with bunting. The field itself throbbed with bustle and activity. Cheering laughter and animated chatter greeted the McKevitts as the trap made slow progress down the crowded road. Brian could barely contain his excitement. In his hand he clasped a half-crown, one of the two Father Pat Trainor had given him. Aunt Bridie said he could spend it at the sports. Hughie parked the trap in a field opposite the sportsfield, unhitched the pony and tied it to the trap wheel.

The sports field was divided in two by a rope barrier which ran the length of the pitch. One half was devoted to various athletic contests – running, long and high jumping, putting the shot over the bar; the other side of the field was cluttered with an amazing collection of stalls – hoop-la, burst-the-balloon, shoot-the-duck and many other "tests of skill," as the stall-minders proclaimed. Interspersed among the stalls were carts packed with fruit, sweets, vegetables and farm produce. James bought an orange for each of the McKevitt party. Brian struggled with the peel until Hughie took the orange from him, slit the peel with his thumbnail and in one deft movement removed the peel in one spiralling loop.

"There," Hughie said, "that's how you skin an Orangeman!"

Brian savoured the juicy flesh of the orange as he sauntered through the crowd, enjoying the atmosphere of gaiety that pervaded the entire field. A cluster of men gathered behind one of the carts. James winked at the others and beckoned them to follow him. Brian wormed his way through the group until he reached a small clearing. A tall thin man with narrow eyes stood before an upturned orange-box which was covered with a green cloth. He shuffled three cards about with great dexterity and kept up a continuous patter.

"Watch her closely, ladies and gentlemen. Watch the lady. Don't watch me! She's much more good-looking! Now you see her, now you don't! Where has she gone? Maybe this young man can tell us!"

Brian suddenly realised that he was the young man. He nervously pointed to the middle card. The thin man turned it face up. It was the Queen of Clubs.

"Well done, young man! You see, ladies and gentlemen. All you have to do is find the lady. They seek her here, they seek her there . . . " He shuffled the cards again and once more invited an onlooker to "find the lady." Once again the onlooker was successful. The thin man seemed delighted.

"I only came here today to give money away, ladies and gentlemen, so place your bets please. Minimum bet one shilling. Who will be first?"

A tousle-haired youth leaped forward. "Begod, I'll give it a try!"

162

He threw a half-crown onto the green cloth.

"Ah, a brave gentleman. Can he find the lady? Watch her now, sir. Watch her!"

He shuffled the cards more quickly this time. The youth pondered for a moment before making his choice.

"Correct! A half-crown to you, sir!"

The thin man produced a gleaming half-crown and placed it on top of the youth's coin. Brian's heart beat faster as he fingered the coin in his pocket.

"Now, who will be next?" Before Brian could move, the figure of Uncle Hughie towered above him and, with a wink towards Brian, he slapped another half-crown on the cloth.

"There y'are!" he shouted expectantly.

"Well done, sir!" The thin man glanced to either side before shuffling the cards. Hughie watched him intently and pounced on a card the moment the shuffling stopped.

"Ah, hard luck, sir!" The thin man turned up the ten of diamonds. "Would you like to try again?"

Hughie withdrew in embarrassment. Brian felt a hand on his shoulder. It was James.

Hughie shook his head in puzzlement as they moved away from the huddle.

"Oh, a fool and his money are easy parted, as the man said," James said lightly, as he cupped his hands around a match to light a cigarette. They wandered around the stalls until an announcement was made about the tug o' war competition.

"That's me," Hughie said and headed off to the other side of the field.

"Hello Brian!"

He recognised Aunt Rose's voice instantly. Hannah stood shyly by Rose's side shading her eyes from the sun. "We thought we might see you here." Rose exchanged greetings with Bridie and James. The twins faced each other.

"Uncle Winston has a tug o' war team. He says they're going to win today," Hannah said.

"Uncle Hughie's in the Cullyboe team. They won it last year." They walked towards the hoop-la stall.

"I have a half-crown to spend," Brian said. "I'm going to have a go." He got six rings for sixpence. The first two were completely off target but he eventually judged the range and the last three hoops landed on top of the bright red cones.

"I won something! I won something!" Brian cried excitedly. The stall owner reached into a box and produced two small articles wrapped in newspaper.

"There y'are now," he said gruffly. Brian tore the paper away to reveal two matching china eggcups with a pink rose design.

"Here," he said on an impulse. "You take one."

"Thanks." She wrapped the cup carefully in the newspaper. They walked on, the adults following a short distance behind.

"I'm going away soon," Brian said.

"Where?"

"Drumcran."

"What's that?"

"It's a convent. I'm going to be living there. And going to school."

"Why?"

"Aunt Bridie says Podgy's too sick and she has to mind him all the time."

"A convent's full of nuns. I wouldn't like that."

"Why?"

"Uncle Winston says the nuns is all the Pope's girlfriends."

"That's silly. Nuns don't get married."

They paused at a stall where Brian bought two sticks of "Peggy's Leg." They walked on in silence.

"Anyway," Hannah suddenly said, "nuns are old and cross and you'll be praying all the time."

The tug o' war competition had reached its final stage. Cullyboe faced Winston Pollock's team, Castleford Red Hands. As the teams made their preparations, the Angelus bell rang in Drumgore village. The Cullyboe team doffed their caps and stood with heads bowed in prayer.

"So well they might pray," Winston said to his colleagues. "They'll need it all! Now boys, for God and Ulster!"

A referee called the two teams to action.

"Best of three pulls, lads. No pulling till I shout *Pull*! And no arguments with the referee!"

The two five-man teams made some last-minute preparations.

164

Resin was applied to hands, belts were tightened, trouser ends were tucked into socks and boots were freed of clay and grass which had built up in the earlier rounds. A long stout rope lay on the grass. Three white handkerchiefs were tied at intervals along the rope.

"Take the rope!" the referee barked. The teams took up their positions, holding the rope loosely. Winston Pollock was the last man on the Red Hands team. Hughie McKevitt stood in the middle of the Cullyboe team.

"Take the strain!" the referee called above a rapidly increasing clamour. The men dug in heels and gripped the rope fiercely. The cries grew louder.

"Come on the Cullies!"

"Up the Hands!"

The referee crouched on his haunches trying to get the middle handkerchief exactly over a whitewashed line between the two teams.

"Ease it a bit, Cullyboe! Another bit. Ah lads, come on. Another bit. Another." The teams cast alternate glances at each other and at the referee. Their facial muscles tautened.

"Pull!"

The shouts of support for either side rose to a deafening crescendo. For a full minute there was stalemate. The middle handkerchief barely moved from above the white line. The advantage swung to and fro as either side's handkerchief moved perilously close to the line. Brian and Hannah stood side by side screaming support for opposite sides.

"Come on, Hughie! Come on, Hughie!"

"Come on, Winston!"

As the front man on the Red Hands team tried to change his grip, a great heave came from the other side and the Red Hands collapsed.

"Cullyboe has it!" the referee called, slapping the ground furiously.

While the roars went up for Cullyboe, both teams took a breather and conferred in a huddle.

"Second pull!" the referee called. Again he had difficulty, this time with the Red Hands, in getting the marker right.

"Pull!"

This time the advantage went to the Red Hands as the Cullyboe men fought to get heel grip on the churned-up grass. One slip and it was over. ˙

"Castleford has it!"

The tension was palpable as the teams took the strain for the final pull. Cries of support for both teams and individuals rent the air as the entire attendance at the sports encircled the tug o' war arena.

"Me life on ye, James McMahon!"

"Pull away, Winston!"

"Drag them to Dundalk, lads!"

"Pull!" The referee's voice cracked above the din. It seemed that the rope might snap in two before either team would cede an inch of ground. The men froze like bronze sculptures in the evening sun, their sinewed arms taut as the rope they pulled, eyes bulging from their sockets as no one dared allow himself a moment's lapse in concentration.

Amid all the din, a strange guttural sound slowly became perceptible. It came from Winston Pollock. The sound grew to a primal roar. It seemed as if he were expelling a demon from deep within him. His temples swelled inordinately as he summoned every ounce of strength into one mighty heave. The result was spectacular. His colleagues collapsed backwards on top of him while the Cullyboe team fell face forward as one and slithered in a heap across the white line.

"Castleford! Castleford has it!" the referee screamed.

There were whoops of delight mingled with polite applause as the crowd dispersed. A breathless Hughie rejoined his family.

"We'll not live that down for a while," he sighed.

"We'd better make for home," Bridie said, knowing Podgy would have expected her home long before now.

"Say goodbye to Hannah, Brian."

Hannah was still gloating over Castleford's victory.

"Uncle Winston pulled all the Taigs by himself."

"They just slipped – "

"They did not. Uncle Winston pulled them all."

"I have to go now."

"Will I see you again soon?"

"I don't know. I'm going to Drumcran."

"Oh yes. To the nuns. You'll be praying all the time. 'Bye."

"Goodbye."

"Thanks for the eggcup," she called.

The Drumgore sports had its reverberations in Cullyboe on the following Sunday when Canon McKenna addressed the congregation.

"I am distressed to learn that there was a women's race at the sports in Drumgore on Thursday last. The teaching of the Church, through our bishops, is quite clear on this. A woman's body is for the bearing of children and as such is to be treated with the greatest respect. It is certainly not intended for public sporting contests – the whole notion of women running in races is shameful and repugnant. It pains me further to learn that a woman from this parish actually won the race. Kathleen Maguire, I say to you that you have demeaned and disgraced your sex and your parish and I forbid you to take part in any such activities again."

He turned away and continued with the Mass.

"Credo in unum Deum, patrem omnipotentem, factorem coeli et terrae – "

Winston Pollock struggled to loosen the collar from his shirt.

"Man dear, we showed them Taigs up, what? The Red Hand'll never be vanquished!"

He ripped the collar away in frustration, threw it on the chest of drawers, took up the glass of whiskey and emptied it in one gulp.

Rose said nothing. She knelt in prayer for a moment before slipping into bed.

"I hope you were prayin' for a child."

Winston tottered as he stepped awkwardly out of his trousers.

"Tell us." He kicked the trousers into a corner. "When are you goin' to give me a child, woman?"

"Keep your voice down, for God's sake," Rose pleaded.

"I'm only askin' a simple wee question. When are you goin' to give me a child?"

There was more menace in his voice now.

"I don't know. It's God's will, surely. And we have a child – "

"She's not our child – and her blood is tainted." He threw a sock in the corner and stood before her in his shirt.

"Please – " Rose whispered.

"You seen the strength of me today. Pulled those Taigs all by

167

myself. You seen it. There's nothing wrong with my blood – or my seed!"

He pulled back the sheet and threw himself across Rose.

"Not a goddamn thing!" he bellowed.

In the adjoining room Hannah Johnston shut her eyes tightly and bit fiercely into the pillow. "I'm not dreaming, Aunt Rose," she thought. "I'm not dreaming."

CHAPTER TWENTY-THREE

TWENTY-TWO, TWENTY-THREE, TWENTY-FOUR . . . TWENTY-FOUR STEPS. There were twenty-four steps to the dormitory in Drumcran. Brian counted them every night. And there were twenty-four panes of glass in the two big windows of the dormitory. Well, twenty-three. One was broken and hadn't been replaced. Mother Caritas said it wouldn't be fixed until the person who broke it owned up. Brian knew who it was and Johnnie Caraher knew who it was, but neither of them would tell. Johnnie was in enough trouble already. He was suffering most because the broken window was at the head of his bed and he shivered in bed each night as an icy blast rushed through the gaping hole. One night he tried to cover the hole with a towel but Mother Caritas had come around after lights out and removed it.

"Justice must be done," Mother Caritas said. Brian's bed was in the corner. He escaped the blast but all twelve boys in the dormitory felt the resulting cold.

"Hey Briny," Johnnie called out. "It's f-fierce cold down here."

"Cover yourself up, head and all," Brian answered in a loud whisper.

"I'd rather be in the cow-byre," came the muffled reply.

There was silence in the dormitory as sleep came to most of the boys. Half an hour later the silence was broken by a resounding thwack, as Mother Caritas swatted at the sleeping figure under the bedclothes with a rolled-up comic she had confiscated in the upper dormitory.

"Johnnie Caraher, what are you doing in there?" She shone a torch into Johnnie's sleepy eyes.

"Sleeping, Mother," he mumbled.

"The way to sleep is with your head on the pillow and your arms over the blankets. Out with them! Out!"

"But it's shockin' cold, Mother –"

"No cheek out of you, now. We know how to remedy the cold, don't we? Don't we?"

The word "remedy" was lost on Johnnie.

"Och, stupid boy. Go to sleep – and keep your arms out!" Brian hastily whipped his arms out over the blankets before the arc of the torchlight swept over his bed.

Johnnie Caraher had come to Drumcran on the same day as Brian Johnston. Since that day he had clung to "Briny," as Johnnie called him. Johnnie was older than Brian by at least four years and had spent the first couple of nights in the upper dormitory with the older boys – the eleven- and twelve-year-olds – but Mother Caritas then decided to move him to Brian's dormitory. Sister Julia had explained to Brian.

"We must be patient with Johnnie. He's a little slow at the books. His mammy has nine other children and his daddy's gone away . . ."

Johnnie sat beside Brian in the classroom, in the chapel and in the dining-hall. He insisted on accompanying Brian everywhere and on helping him with his daily tasks.

"You and me is goin' out the walks, Briny," he would say. "You and me is goin' to polish the altar rails, Briny."

From the day he entered Drumcran, Johnnie was rarely out of trouble. He refused to eat the porridge for breakfast until Mother Caritas stood over him and made him eat it.

"Good nourishing food mustn't be wasted. Eat up!"

Johnnie screwed up his face and sloshed the stodgy gruel slowly into his mouth.

"There!" Mother Caritas said. "That wasn't too difficult, was it?"

She was just about to turn away when Johnnie vomited everything up. Brian couldn't bear to look at the mess nor at Johnnie when Mother Caritas ordered him to eat it again. Johnnie began to cry but Mother Caritas would hear no excuses.

"Of course you can eat porridge without sugar. It's good self-denial!"

Brian's own stomach heaved at the sound of Johnnie forcing the vomit down his throat. It stayed down this time.

Sister Julia summoned the two boys on their way from evening prayers. They followed her into a dark alcove.

"Here!" She pressed an envelope into Brian's hand. It was full of sugar.

"I don't take sugar in my tea, so put it in Johnnie's porridge. And here!" She pressed an orange into each boy's hand. "Good for you!" she smiled and shooed them down the corridor. Later Brian remembered how Hughie had shown him how to peel an orange. He took Johnnie's orange and made a crude attempt to follow Hughie's example.

"That's how you skin an Orangeman!" he laughed.

Johnnie again incurred the wrath of Mother Caritas when he occasionally wet his bed.

"A big boy like you. Shame on you!"

Johnnie was forced to stand by his bed while Mother Caritas draped the stained and still wet sheet over his head.

"We'll have no more of this nonsense!" she said. But it did happen again and each time he stood for fifteen minutes under the wet sheet. "I think we'll have to put you in the cow-byre," she said one night.

The next night Johnnie disappeared. There was a major alarm until he was discovered sleeping in the manger of an empty stall in the cow-byre.

"It was cushy and warm there, Briny," he said as Brian sprinkled some sugar over Johnnie's porridge at breakfast.

And now there was the broken window. Three of the big boys from the upper dormitory had sneaked into Brian's dormitory before evening prayers.

"Hey Johnnie, show us your mickey!" the boy called Malachy sniggered. "We hear you have a big one!" The three pinned Johnnie on his bed and attempted to pull down his trousers. Johnnie accepted the prank as a game and laughed and giggled as the big boys tumbled him over. In his exuberance Malachy put his elbow through the window, whereupon the three scuttled out of sight in an instant.

Brian lay awake listening to Johnnie shivering and turning to and fro in his bed. He decided he would tell Mother Caritas. Not what really happened. He knew Malachy would come back if he did that.

171

"And how did it happen?" Mother Caritas asked.

"We were just – playing."

"And why has it taken until now to tell me?"

"I – I was afraid, Mother," he whispered.

"Speak up!"

"I was afraid – Mother."

"You should never be afraid of the truth, Brian. And you should stay clear of Johnnie Caraher. I don't know what we're going to do with that child. Now, an hour after school in the chapel for the next three days. And be a good boy from now on."

The chapel was warm at least. He sat beside a statue of the Blessed Virgin. She was standing on top of a snake, squashing its head. He made a face at the snake. It snarled back at him.

Ssss! it said. Ssss! Ssss!

Sister. Hannah wouldn't be in a chapel. She would be feeding her calf, maybe. Or helping Aunt Rose. Or just playing in front of the fire . . .

Underneath the snake were the words *Immaculate Mother.* What did "Immaculate" mean? What did "Mother" mean? Mother Caritas. He was afraid of Mother Caritas. Johnnie had a mother but Johnnie was here.

"Briny! Briny! There's a man fixing the window!" Johnnie slid in beside him. "Why are you in here, Briny?"

"Mother Caritas sent me in – for – for being bold."

"I'll stay with you, Briny."

"No, no. You can't. Mother Caritas said I was to be on my own."

"I'll stay with you, Briny."

"You can't. You can't. You have to go. Mother Caritas will be coming in soon. Go out to the field."

"Don't want to. Big Malachy's out there. Want to stay with you, Briny."

"Well, you'll have to hide – in the confession box – and stay quiet."

Johnnie did as he was told. Minutes later he began to hum a song to himself.

"Shush, Johnnie!" There was silence again. Mother Caritas looked in.

"Good boy! Ask God for forgiveness now," she said. Brian glanced at the confessional. He would have to tell Father O'Hagan in confession about the lie he told Mother Caritas . . .

The frosty air nipped his cheeks and ears but Brian was happy. Hughie had come to collect him in the pony and trap. He was going home for Christmas. Johnnie ran alongside the trap as it moved off down the avenue.

"I'll see you after Christmas, Briny," he said. Brian felt sorry for Johnnie who would be spending Christmas in Drumcran. When the trap reached the road, Johnnie stood by the wrought-iron gate, waving until the trap disappeared from view.

"I'll see you, Briny." Johnnie's words echoed in the chill December air.

Aunt Bridie ran to greet Brian as he descended from the trap. She hugged him warmly.

"Och, I wouldn't know this big man," she cried. "I missed you round the place." She shepherded him into the kitchen. Podgy was seated in his chair by the range.

"Begod, you're home," he grunted.

"Hello, Uncle Pod – " Brian's eyes fell on the blanket draped across Podgy's lap. He stood staring at how the blanket suddenly fell to the floor on one side. Podgy's left leg was missing.

"Come on, Brian." Bridie urged him towards the door. "I have a surprise for you upstairs."

He was still looking back at Podgy as he went up the stairs. Bridie waited until they reached the landing before giving the explanation his eyes demanded.

"Podgy has been very sick with his leg," she said quietly. "It was all poisoned. He had to go to hospital to have it taken off." She gave a choking sigh. Brian felt the tears welling in his eyes.

"Did it hurt him?" he asked.

"A bit." Bridie ruffled the boy's hair. "Just – just don't pass any remarks on it." She drew a deep breath. "Now! Here's your surprise!" She took him by the hand to a door that he had never seen open before. She stood at the threshold. "Go on in!" she whispered.

It was a small bedroom, newly decorated and bathed in the glow of a tall oil-lamp that sat atop a chest of drawers. It was sparsely furnished with a bed, chair, chest and wash-stand but its colours were warm and welcoming. Brian turned in surprise.

"Is this – ?"

"Your very own room! Do you like it?"

"Oh yes! Thank you, Aunt Bridie!" He bounded into the room, his face beaming with delight. He ran his hand along the coverlet on the bed. He wanted to touch every part of his room.

"And look what I found!" Bridie announced, taking up a framed photograph that stood beneath the lamp. Brian studied the photograph closely. Two young men and two young women stood smiling nervously in a field.

"I couldn't find that for years till Josie came across it up in Drominarney. I mind the day the Yankee Kerley came down to the flaxfield showing off the camera he brought home from America – "

"But who are they?"

"Don't tell me you don't know your Aunt Bridie!" she laughed, pointing to the dark-haired girl on the left. "And that's Hughie! And your Uncle Pete that died of the weakness. Always laughing!" Brian followed her finger along the photograph to the fair-haired girl who looked shyly towards the ground.

"And that is Sarah, your mother."

Brian lay between the warm flannel sheets that night feeling snug and happy. He luxuriated in the comfort and privacy of his own room. There would be no Mother Caritas, no Johnnie Caraher, no noise, no smells. He thought of Johnnie, probably on his own in the dormitory. "I'll see you, Briny," the words rang in Brian's head.

The lamp was turned low. He reached across and took the photograph from the top of the chest. He gazed intently at the fair-haired woman to the right of the group. Mother. Mother Caritas. Immaculate Mother . . .

They were stacking sheaves of oats. Everyone was happy. His mother picked him up and swung him round and round. Suddenly there was a wild whoop behind them as a rat ran from a nearby stack. Hughie swiped at it with the pitchfork. The rat spun in the air and, as it fell, Hughie stomped on it with his boot. Brian was still being swung around and around. He was growing dizzy. The spinning stopped. It wasn't Hughie any more but a one-legged Mother Caritas who was squashing Johnnie Caraher under her foot. "I'll see you, Briny," Johnnie kept calling. "I'll see you, Briny . . . "

174

"It's all right, Brian. It was only a dream." Bridie's voice was soft and reassuring. She replaced the photograph on the chest of drawers and eased his head back on the pillow.

On Christmas morning Brian woke to find a football, a storybook and a box of sweets at the foot of his bed. Santy Claus had kept his word. Brian and Johnnie could play endless games of football now, although he did not want to think of going back to Drumcran yet.

Podgy had grown more grumpy and rarely used the crutches he had been given. Instead there were constant demands on Bridie's attention.

"Am I goin' to get any tay this day?"

"Will ye fix these curse-of-God cushions?"

"I need the bottle!"

The "bottle" was a large jamjar into which Podgy would urinate. Bridie would then have to empty its contents in the ditch. Brian kept out of Podgy's way as much as he could, staying in the shop with Bridie or playing football with Hanratty in the yard.

On St Stephen's Day he walked with Bridie to Drominarney. He was given a hearty welcome by Hughie, James and Josie and spent a long time playing with Peter, rolling the ball to and fro across the kitchen floor. They had lemonade and Christmas cake and Josie played a few tunes on the fiddle.

As soon as Bridie put her hand on the latch on their return, there was an angry roar from Podgy.

"Where did ye put the curse-of-God bottle? I'm after destroying myself here. A nice way to be left!"

The stench of urine and the pool that seeped slowly into the crevices of the flagged floor told the rest of the story.

"Go on up to your room, Brian," Bridie said quietly as she hung her coat on the back of the kitchen door.

He lay on his bed and thumbed through the *Best Stories for Boys*, while the sounds of heated argument came from the kitchen below.

A week later Hughie called to leave him back to Drumcran. Podgy called a gruff "Good luck, young fella" to him. Bridie held him in her arms for a long time.

"Don't want to go," he snuffled into her apron.

"It's best," she whispered as he stepped into the trap. She kissed him and stood in the raw January evening waving to him until the trap disappeared into the gloom.

Johnnie Caraher was swinging one-handed from the wrought-iron gate of Drumcran. His face lit up on Brian's arrival.

"You were a long time, Briny! I came down every day to see if you were coming."

CHAPTER TWENTY-FOUR

Master McCartan sat with his back to a glowing fire. On the desk before him a pile of opened copybooks lay on one side and a pile of closed ones on the other. He sighed frequently as he waded through the homework corrections. He detested correction. The same mistakes. The same sloppy, half-done work every year for thirty-five years. He took up another copybook and gave an angry snort.

"Caraher. Where's that eejit?"

The boys stopped in the middle of their transcription.

"Here, Master." Johnnie smiled up at him from the bench at the back of the room.

"Come up here and I'll give you something to smile about!"

He shoved the chair back awkwardly and held Johnnie's copy aloft.

"Pens down, boys! It's not often you'll get a chance to see a piece of art like this!" There were sniggers from the front benches.

"Draw a map of France showing the chief cities and rivers. That was the exercise. A simple enough task, I would have thought. And this –" he held the copybook at different angles for all to see – "this is how Jean Toulouse Caraher visualises that noble country, the birthplace of democracy!"

The "map" was a smudged roughly rectangular outline, punctuated with several large blobs from which rivulets of ink had run before the blots had dried.

"Looks more like a burst bag of blackberries to me, boys." The class laughed openly now. "What is the capital of France, Caraher?"

177

Johnnie shrugged his shoulders and smiled.

Brian wanted to whisper the answer to him but he was too far away.

"Tell him, Murphy!"

"Paris, Master."

"And what river is Paris on?"

"The Seine, Master."

"That's right. Paris on the Seine and" – he fluttered Johnnie's copybook – "Caraher in-sane! Isn't that a good one, boys? Paris on Seine and Caraher insane!"

The boys laughed as he had prompted them to do. He flung the copybook into the fire.

"Did you ever see Paris, Caraher, in your many travels?"

"No, Master." Johnnie was still grinning.

"Well come here – and I'll show you Paris."

He turned Johnnie to face the class, stood behind him and firmly gripped his two ears before lifting him off the ground. Johnnie squealed in pain. "D'ye see it now, Caraher? D'ye?"

Master McCartan's face grew redder as his arms felt the strain of lifting Johnnie's body. Suddenly he let go and Johnnie fell to the ground, howling and cupping his ears in his hands in a vain attempt to ease the searing pain. The master fumbled angrily at his desk until he found the rod, a long sally stick that whirred in the air as he brought it down across Johnnie's legs.

"Get up! Get up and stop your whinging! Right hand!"

Johnnie extended his left hand nervously.

"God in Heaven. Can anyone be so stupid?" The master whipped the stick upwards into Johnnie's knuckles and grabbed his right hand. Six stinging blows followed. Brian winced at each one and felt the tears well up in his eyes.

"Left hand!" Six more blows.

"And don't – produce – homework – like that – ever – again!"

The master panted as he punctuated each word with a blow. Brian felt his eyes hot and wet. He covered his face with his hands. Through his fingers he could see the crouched figure of Johnnie, squeezing his hands under his armpits and, beyond, the charred pages of his copybook curling in the leaping flames.

When classes were finished for the day, Master McCartan kept back four boys including Brian. These were the "scholarship boys" who would sit an examination at the end of June to compete for scholarships to secondary school.

"Now boys, I've been going through the old geography papers and I've compiled a question for you. Atlases out, boys! Write a brief note on the location of five of the following: The Potteries, Marseilles, the Caspian Sea, the Kiel Canal, Elba, Omsk, Spitsbergen . . . Spits Bergen – sounds like one of Al Capone's sidekicks. Spits Bergen. Isn't that a good one, boys. Man dear, I'm at the top of my form today, what? Paris on Seine and Caraher insane, ses he. I wonder what Caraher would make of Spitsbergen. Maybe we should send him there, what do you think, Brian?"

Brian jumped at the mention of his name.

"S-sorry, sir – "

"I was saying, should we send Caraher to Spitsbergen."

Brian chewed on the end of his pencil.

"I-I don't know, sir. Haven't found it yet," he stammered.

"Well hurry up, man. Time is of the essence!"

Brian's mind was far from Spitsbergen. He was thinking of the letter that he had received that morning. He had read it several times during the day.

June 15th 1937

My dear Brian,

I hope you are keeping well. Things are not too great here. Podgy has had to go into hospital. His other leg has got very bad and I can't manage to look after him properly. He is a very sick man. I want you to pray specially for him as I know you are a good boy at the prayers.

I'm writing this so you won't get a shock if Podgy is not here when you come home in a couple of weeks. And to wish you well in the scholarship exam. I know you will do your best. God's will be done in all things.

God bless you,

Your loving Aunt Bridie.

Brian slipped into the chapel that evening after play. Inevitably Johnnie followed him.

179

"What are ye prayin' for, Briny?"

"My Uncle Podgy. I think he's dying. Aunt Bridie said to pray for him."

"D'ye know what, Briny?"

"What?"

"I'm goin' to pray that Master McCartan is dyin'. Then he won't be able to beat me any more."

"Shush, Johnnie!"

"Shush, Briny!"

There was a brief silence.

"Briny?"

"What is it now?"

"You're goin' home soon, for good."

"Yeah."

"Can I come with you – for a wee while?"

"I don't know, Johnnie. I'll see."

Podgy Kernan died in hospital on June 26th 1937. When Bridie saw Owen Brady cycle towards her at the unusual hour of four o'clock in the afternoon, she knew Podgy was gone.

"I'm terrible sorry, Bridie," Owen said, handing her the telegram. "Poor ould Podgy, he wasn't the worst."

"Aye." Bridie gave a deep sigh. "His suffering is over, at least."

"That's true. Tell us, can I do anything for you, Bridie?"

"No, thanks. This day is not unexpected."

When Owen had gone, she went inside, and drew the blinds on every window in the house. She sat for a while in the darkened kitchen, took out her rosary beads and prayed for her husband's soul. Memories came to her unbidden, all of them warm and cherished and all of them at Drominarney. Podgy's kindness on the night of Sarah's death, the cot he had made for Brian, the summerhouse he had given for Pete. Bridie wiped her eyes, took up her cardigan and made for the back door. She would cycle up to Drominarney. Hughie and James would help her now. On her way through the scullery she knocked something just inside the door. It was Podgy's "bottle." She picked it up, walked across the yard and hurled the bottle with all her strength against the wall of the dung-heap.

Podgy's funeral was one of the largest seen in Cullyboe for many years. Before the final blessing, the altar boys placed a table covered with a white linen tablecloth before the altar. Canon McKenna stood at the table, flanked by Hughie McKevitt and a cousin of Podgy's. As the mourners filed forward with their Mass offerings, Hughie and Podgy's cousin identified them in turn for the Canon. The altar boys gathered the tablecloth by the four corners and removed it to the sacristy.

Two topics dominated conversation in the pubs of Cullyboe that evening. The size of Podgy's offerings and the impending general election.

"I never seen a take as big as that before."

"Begod, the Canon'll have a new car on the strength of it."

"Dev could do with it for his election fund!"

"He wouldn't get a bleddy penny of mine."

"I seen him over in Dundalk last week. He made a brave speech, telling us all to vote for him and his constitution."

"He can blather all he likes," Owen Brady muttered. "He'll get no votes in our house. Outlawing the IRA like that. His own brothers in blood. He can take a runnin' jump at himself."

Brian and Johnnie sat on the steps of Drumcran House. Brian thought deeply about Aunt Bridie's news. Podgy was dead and buried. Brian couldn't say he would miss Podgy. He remembered his clammy hands and how much he had been in fear of Podgy over the last few years. But why hadn't Bridie brought him home from Drumcran for the funeral?

"Because you were in the middle of your scholarship exam," Bridie had said. "That's important for your future."

"D'ye think she'll take me, Briny?" Johnnie interrupted his thoughts.

"Don't know. Hope so," Brian said.

Bridie was discussing with Mother Caritas the possibility of Johnnie spending some time with Brian.

"She's talkin' a long time, Briny," Johnnie said, looking anxiously around. "Will we go over to the window and listen?"

Just then the door opened and Bridie emerged from the convent parlour.

"Get your things, Johnnie. You're coming on your holidays!" Johnnie gave a little whoop and raced off to the dormitory.

"It's only for a week or two," Mother Caritas called after him. "And if you cause any trouble, John Caraher, you'll be back here before you know it." She turned to Brian. "I'll be expecting great things of you, Brian Johnston. Goodbye and God go with you, child."

Brian took her extended hand but his hand slipped instantly from the moistness of hers.

Bridie enjoyed the excited chatter of the two boys. They laughed and sang and waved cheerily to anyone who happened along the road. It would be good for Brian to have a friend at home, if only for a couple of weeks. It would be good to hear laughter replace silence, fun replace pain . . .

In the ensuing weeks the boys were inseparable, roaming through the fields aimlessly, helping Hughie save the hay and spending long sunny evenings splashing about in the river. On one particular golden evening they lay on the river bank, exhausted after their exertions.

"D'ye know something, Briny?" Johnnie said as he chewed determinedly on a long stem of grass. "I'm going to be a millionaire when I'm big. I'll live in a big, big house and I'll drive a big, big car. Vroom! I'll bring you for drives, Briny!"

"Where are you going to get all the money?"

"I'll have a factory and we'll make pound notes, a hundred of them, every day."

Brian laughed, plucked a dandelion and threw it at Johnnie.

"What'll you be, Briny?"

"Don't know."

"I'd say you'll be a teacher – or a priest!"

Brian laughed even more heartily and sprang to his feet.

"Race you back to the house!" he cried.

"I've got good news for you, Johnnie," Bridie said over tea. "Mother Caritas says you can stay another two weeks but you have to go working for a farmer in Inniskeen then." The boys cheered in unison. Their enjoyment continued unabated, playing tricks on Hanratty, chasing Fox's donkey which defied all Johnnie's efforts to ride it and spending whole days in the company of Thomas, Charles,

Kathleen and Peter Joseph, Brian's cousins in Drominarney. There was great excitement one afternoon when Uncle James came back from Castleblayney and struggled up to the loft with a large box. The children watched with growing curiosity as James unpacked a mysterious machine.

"What is it, Uncle James?"

"It's a magic box, Thomas. Just wait till you see what it can do!"

He spent a long time working at the back of the magic box before eventually warning the children to stand well back. They edged slowly towards the door as James turned a dial. There was an expectant silence until a little red light began to glow in one corner of the box and a voice began to speak from within. The children stood open-mouthed until Thomas began to cry. He burst through the spellbound group, dashed down the steps and raced across the street into the kitchen, his round face etched with fear.

"Mammy! Mammy! Uncle James has a wee bogeyman in a magic box!"

The wireless had come to Drominarney.

On their way home from Drominarney a car pulled up alongside Brian and Johnnie.

"Who are you two boys?" a gruff voice spoke from within.

"I'm Johnnie and this is Briny. Who are you?"

"I'm Canon McKenna. Kindly address me as 'Your reverence,' foolish boy! I know you." He poked a finger towards Brian. "You're young Johnston, aren't you?"

"Yes, Canon."

"Yes, YOUR REVERENCE! What are they teaching you boys nowadays? Tell me. What is sanctifying grace and where do we get it?"

"Briny's Aunt Bridie has a shop. You'll probably get it there," Johnnie replied before Brian could open his mouth.

"Achh!" The car sped off down the road, enveloped in a cloud of dust.

On the last day of July, Johnnie left to start work in Inniskeen. Charlie McMahon gave him a lift to Castleblayney. Bridie and Brian watched and waved as the trap swayed down the road. The slight figure clutching a tattered canvas bag called back as the trap turned a bend in the road – "I'll see ye, Briny!"

"God go with him," Bridie said. "I wonder when we'll see him again."

Two days later Owen Brady delivered two letters.

"It's either a feast or a famine," Bridie said. "Will you take a glass of something?"

"Begod, a bottle of stout wouldn't go astray. I'm near the end of my run anyway. 'Tis thirsty oul' weather."

Bridie opened the bottle of stout.

"Don't bother with a glass," Owen said. "Only makin' work for yourself!"

He put the bottle to his lips and took a long draught. Bridie glanced at the letters. One was addressed to Master Brian Johnston. She tore it open nervously and read the letter. It began:

Dear Brian,

I am pleased to inform you –

She ran to the back door of the shop shouting Brian's name.

"Mother of Christ! Not bad news, I hope," Owen called.

"No! No!"

Brian came running from the bottling shed.

"It's the scholarship! You got the scholarship to St Felim's!" She thrust the letter into his hand and ruffled his hair.

"I'm real proud of you!"

"Good gossoon!" Owen said as he wiped his lips with the back of his hand. "The schoolin'll be the makin' of ye. I'll see you, Bridie. Thanks for the bottle!"

Brian read the letter several times. He would start in St Felim's on September 4th. Bridie read his mind.

"I'm going to lose you again, just when I was enjoying having you around the place. Never mind. This is great news. You're a good boy, Brian. Your mother would be proud of you. Come on. We'll have a lemonade to celebrate – and a packet of fancy biscuits!"

Amid all the excitement of Brian's news Bridie forgot about the second letter. It was late afternoon before she came across it again. It was addressed to her and was neatly typed. "I hope this is as good news as the first one," she mused as she unfolded the letter. She read with growing disbelief.

William Plunkett, Solicitor
Main St
Cavan
August 1st 1937

Dear Mrs Kernan,

I am instructed by my client, Joseph Kernan, a cousin of your late husband Peter, to notify you that there is a debt of £3,000 outstanding to my client. This represents a loan given to your late husband by William Kernan, father of my client, in 1920 towards the purchase of the premises you now inhabit. The debt includes the original loan plus interest accruing since 1920.

My client has lodged with me a promissory note confirming the said loan. He now demands payment of the outstanding debt. Alternatively, he would be prepared to accept a share in his cousin's premises in lieu of payment.

Perhaps you would favour me with your intention of dealing with this matter as expediently as possible.

I remain,

Yours sincerely,

William Plunkett, Solr.

CHAPTER TWENTY-FIVE

"Salt – mustard – cayenne pepper," the girls chanted slowly as they swung the skipping-rope in time with their rhyme.

"Come on, Hannah! Jump in and stop your dreamin'!" Hazel called. "The parade will be here soon."

Hannah resented her cousin's bossiness. Hazel was a year older than her and liked to affect a big sister attitude to her "country cousin" as she called her. Each July Hannah stayed with her second cousin in Belfast and in August Hazel returned the visit in Sloe Hill. It was not an arrangement that Hannah looked forward to but Aunt Rose insisted that it was important to know one's extended family. For three years now this arrangement had occupied Hannah's school holidays.

It was the morning of the twelfth of July. The bunting and the Union Jacks hung limply from lampposts and windows. The grey misty morning had enticed only a few stragglers onto the pavement in anticipation of the Orange parade. Consequently, the three girls – Hannah, Hazel and her friend Marjorie, a slight bespectacled girl of ten years, younger still than the cousins and even more cowed by Hazel's domineering ways – had a long stretch of pavement to themselves for skipping.

Hannah reached through the railings of a front garden and withdrew a dandelion in seed.

"Oh all right, Hannah," Hazel sighed, "let's see who you're going to marry."

Hannah began to blow away the seeds while the other two chanted:

"Tinker, tailor, soldier, spy
Rich man, poor man,
Beggarman or thief!"

"Maybe I'll not marry anyone," Hannah said.

"Wait!" Hazel cried. "Keep puffin'. I know another verse.

Chinaman, African,
Jap-an-ese;
A clergyman, a rabbi
A papish priest!"

"A papish priest! Hannah's goin' to marry a papish priest!" Hazel jeered.

"Am not. I was finished at 'clergyman.' You just kept going!"

"Och, come on. You take the rope, Hannah, and I'll jump in!"

Hannah swung the rope absent-mindedly with Marjorie.

A clergyman. A clergyman . . .

Mr Barden had taken Sunday School on the previous Sunday. He was new to the parish, his first posting since his recent ordination. Hannah was totally captivated by his warmth, his friendly smile and his soft voice. He had taught them a verse which Hannah constantly recited to herself.

Who is on the Lord's side? Who will serve the King?
Who will be His helpers other lives to bring?
Who will leave the world's side? Who will face the foe?
Who is on the Lord's side? Who for Him will go?

She would recite it for Mr Barden next Sunday. She was on the Lord's side. She was on Mr Barden's side. He could only be in his twenties. When she would be in her twenties, he would be in his thirties. That was all right. As long as he waited . . .

"Apple, jelly, jam-tart
Tell us the name
Of your sweet-heart
Mister B – Mister A – Mister R
Mister D – Mister E – Mister N"

Hazel exploded into laughter at her own rhyme. Marjorie giggled shyly.

187

"I hate you, Hazel Taylor," Hannah cried. She threw the skipping-rope into the gutter and ran away up the street. Just then the first band in the Orange parade rounded the corner at the top of the street. It was a signal for doors to open and, within minutes, the pavement on either side was lined with eager spectators, mostly women and children. Hannah found herself being pushed forward onto the street itself. The crowd cheered in response to frenzied drumming of three young men at the rear of the band. Then came a huge banner depicting King William on a magnificent white horse with the legend:

Remember 1690 – For God and King Billy underneath. It was borne aloft by two men who were relieved by two others every hundred yards or so. Behind the banner came a phalanx of bowler-hatted men, many of them wearing sashes and bearing rolled-up umbrellas. They marched proudly, looking straight ahead and seeming almost unconscious of the noisy crowd. In the front row, Winston Pollock strode out purposefully. Hannah waved and called out to him but there was no response. A young man pranced up and down alongside the parade, twirling a baton and tossing it to a prodigious height before catching it again to raucous appreciation from the onlookers. As the tail-end of the parade approached, Hannah was about to turn away when she noticed a familiar figure in the second-last row. It was Mr Barden! Unlike the others, he was smiling shyly towards the crowd and as he drew level with Hannah he gave her a special wave of recognition. Hannah blushed as she waved back. In a moment he was swallowed from her view by a mass of young people who fell in at the end of the parade.

"They're away to the field now," a stout woman behind her said. Hannah didn't know where the field was but she didn't care either.

"Now, Miss Hazel Taylor," she thought. "I hope you saw that! Mr Barden waved to ME! I'll bet he didn't even see you! 'Who is on the Lord's side? Who will serve the King?'" she hummed to herself as the crowd thinned away.

Bridie negotiated the pony and trap into the yard of Markey's shop.

"Now, Brian, you and me are going shopping while Uncle James goes about his business," she said while tethering the pony. James lit a cigarette. "And we'll meet in the Central for a big dinner at one o'clock, so don't be late."

He winked at Brian and hobbled down the yard, leaving a trail of smoke behind him. Inside Markey's, Bridie worked her way methodically through a shopping list for Brian – trousers, shirts, pyjamas, socks, jumpers, shoes. He looked away in embarrassment when a shop assistant held a pair of long johns up against him for size.

"Do I have to wear them?" he muttered to Bridie.

"Come the winter, you'll be glad of them," she replied aloud. "Won't he?" she addressed the assistant.

"Oh, them's the only boys for the could weather," he said cheerily.

Brian cringed. He gazed around the vast shop in an attempt to show his indifference to the matter. The large signs intrigued him. Haberdashery, Millinery, Linens, Ladies' Fashions, Lingerie . . . The most fascinating aspect of the shop was the payment of bills. Bridie tendered two notes to the assistant who promptly folded them with the bill and despatched them in a container along a pulley-operated wire to a mysterious room that seemed to be suspended from the roof of the shop. A few minutes later the container returned at great speed. The assistant opened it and presented Bridie with the bill and her change.

"Come on, Brian. You're in a total dream there," she said. He followed her down the aisles, still looking back to the mysterious room, wondering what strange beings worked there. "We just have to get your bed-linen and we're away then," Bridie said, fingering some curtain material as she made her way to the Linens Department.

They sat in the Central Hotel for ten minutes before James came along.

"We were just on for going home," Bridie joked.

"Och, Tommy Murray would talk the hind legs off an ass," James said as he patted his pockets in search of his cigarettes.

"What's 'News of the World' saying now?" Bridie asked as she read through the menu.

"He's predicting another General Election – says de Valera is living dangerously with no overall majority – and he says there'll be a war before long with the way the Germans are carrying on."

"God bless us but he's full of cheer. We'll all have the roast chicken and ham," Bridie said to the waitress. "Has he no good news at all?"

"And plenty of spuds for this growing man," James called after the waitress. "Well I'll tell you at three o'clock," James said in answer to Bridie's question. "If a certain Red Prince wins the big race at Ascot." He winked at Brian.

"The two of you are well met," Bridie sighed. "Did you get any business done at all?"

"Of course." He reached into his pocket. "And I got something for Brian too."

He handed the package to Brian who opened it with anticipation. It was a prayer-book. *The Garden of the Soul* by Bishop Challoner.

"You'll not go far wrong if you abide by that," James said. Brian leafed through the pages. There were Morning and Evening Prayers, Devotions for Mass, visits to the Blessed Sacrament, Prayers for Confession and Communion.

"Thanks, Uncle James," Brian whispered.

"If you remember the prayers, Brian, everything else will fall into place," James said. "And when we get home I'll make you a nice leather case for that book. Now where's that dinner? I'm starving!"

James was unusually quiet on the way home. Bridie shook the reins at the pony. "Hup!" she called. "You're nearly, *nearly* as slow as Red Prince! And I remember someone telling Hughie a fool and his money are easy parted." James accepted the barb with equanimity.

"Let's hope Tommy will be wrong in all his predictions," he said.

A steady drizzle depleted the attendance at Drumgore Sports on August 15th. Bridie decided not to go but Brian was relieved when Hughie and Peter Joseph arrived in the trap. He scoured the field in search of Hannah and had almost given up hope of seeing her when she appeared late in the afternoon with a girl of her own age.

"This is Hazel, my second cousin," Hannah said. "She's from Belfast."

"This is Peter – my – our first cousin," Brian replied awkwardly.

"Aunt Rose didn't want us to come because of the rain, but we persuaded her. Hazel wanted to see you!"

"I did not, Hannah Johnston. I'm not speaking to you ever again." Hazel affected hurt but stayed close to her cousins.

"Are you still with the nuns?" Hannah asked.

"No. I'm away to boarding school in September. St Felim's. I got a scholarship."

"Are you going to be a – priest?" Hazel asked.

"No!" Brian laughed.

"He's going to be a scientist, a big inventor – that's what he told me!" Peter Joseph jibed. Brian jabbed at his cousin playfully.

"I'm going to Miss Dagg's Academy," Hannah announced.

"What's that?" the boys chorused.

"It's a school – in Belfast. I'll be staying with this one – unfortunately." She nodded towards Hazel.

"Hmph! Nobody's making you stay anywhere," Hazel countered. "You can stay with Mr Barden if you like – "

"Hazel Taylor!" Hannah flushed with anger and embarrassment.

"He's our new rector. Only a baby. Hannah fancies him – "

Hannah felt tears well up in her eyes. She turned away abruptly.

"I'm going home," she whispered. "Youse can do what you like." She disappeared through the crowd in an instant.

She cycled home furiously, wanting to put as much distance as she could between herself and Hazel. Tears blurred her vision and on a couple of occasions the bicycle mounted the grass verge. She turned into the gates of Sloe Hill, threw her bicycle against the hedge, hurried through the kitchen and up the stairs to her room. She lay on her bed sobbing quietly.

"What is it, Hannah?" Aunt Rose spoke quietly from the doorway.

"Nothing."

"You're upset. I saw the way you came in."

"It's nothing."

Rose sat at the edge of the bed.

"Hannah – "

"I'm not going to school in Belfast. I hate Hazel."

"Why?"

"She's always teasing me, making a show of me."

Rose put her arm around her niece and raised her to a seated position.

"That's just Hazel's way. She doesn't mean any harm – "

"I hate her."

"No, Hannah, no." Rose held the girl's quivering body close to her. "I'll talk with Hazel – "

"No!"

"With her mother, then."

She rocked Hannah gently to and fro.

"It'll be all right. You'll see. And you'll make lots of new friends in the Academy." The sleeve of Rose's cardigan had ridden up above her elbow. Hannah noticed the bruising as she wiped her eyes.

"Aunt Rose?"

"Yes, Hannah."

"Does Uncle Winston – beat you?" She gulped the last words.

"Hannah!" Rose jerked backwards, hastily pulling down her sleeve. "Hannah! I'm surprised at you. Winston is a good man. He provides for us all. That bruise – it's just where – the roan cow jammed me against the wall – at milking-time. That's all. Don't you remember?"

Hannah shook her head.

"Hannah. Don't ever – ever – say things like that again!" Her tone had changed completely. Hannah nodded meekly. "Now have a wee rest while you're here and leave Hazel to me!"

Rose withdrew quickly from the room. Hannah lay back on the bed. She was confused. She had not meant to upset Aunt Rose. She reached into the bedside locker and felt under the clothes until she found the hymn book. She opened the book and took out a neatly folded leaflet. She read the printed verses slowly.

Little to me it matters
Whither my feet are led,
If in the burning desert
Or the pastures green I'm fed;
Whether the storm or sunshine
Be in the path I take
For my hand is in Thine, my Father,
Thou will not thy child forsake.

And it shall not cause me sorrow
Though the path be steep and rough
I am Thine, Thine own forever
And that shall be joy enough.
Thine is the care, my Father,
The work of providing Thine
Only the trust and pleasure
And the calm content are mine.

At the bottom, in neatly rounded handwriting was the inscription:
For Hannah
The Lord is always with us.
William Barden

CHAPTER TWENTY-SIX

BRIDIE FOLDED THE LETTER AND SLID IT BACK INTO THE ENVELOPE. IT WAS the third letter in as many months from William Plunkett, Solicitor. A reply to the first letter from Hugh McGribben, Podgy's solicitor, had had no effect. Podgy's cousin was still demanding three thousand pounds "as a matter of extreme urgency," as this latest letter put it. Three thousand pounds. She shook her head slowly. She might as well give him the whole place, lock, stock and barrel. She reached up to place the letter behind the tarnished plaque on the top shelf behind the bar – a plaque that commemorated Podgy's sole success when he had ventured into greyhound racing many years ago.

"Come right out from behind that bar, honey, and let me give you the full of my arms!" The slight American accent failed to mask the unmistakable Monaghan burr. Bridie paused momentarily to look in the smoke-hazed John Player mirror at the figure that stood, arms outstretched, in the middle of the floor on the other side of the bar. She turned to face the tall slim figure almost totally enveloped in a tightly belted crombie coat, the cheeky grin peering from under the brim of a trilby hat.

"Jimmy McCabe, I declare to God!"

"One and the same."

"One and not the same! Look at the cut of you. Straight out of millionaire's row!"

"Aw shucks, Bridie McKevitt, you know I'm just a simple country boy. Now are you going to stand there looking at me or am I going to have to jump that bar and sweep you off your pretty little feet?"

Bridie blushed with embarrassment as she noticed Hanratty peer in through the back window of the bar. She waved him away with one hand as she lifted the bar flap with the other. She stood awkwardly against the bar for a moment before Jimmy came and swept her into his arms. Her head rested between the collar-folds of his overcoat. She clasped her arms tightly about him, shut her eyes and inhaled the strange heady scent that his body exuded. When had she last savoured the warmth of a man's body, felt his breath on her forehead? When? When?

She shuddered as she remembered Podgy's awkward fumblings during the early weeks of their marriage. Jimmy nuzzled deeper into her hair. She could feel his breath on her scalp. She shut her eyes even more tightly. A warm June evening. The scent of new-mown hay, hanging like a great sheet of gauze all over the valley. Voices. Echoing laughter. Sarah's laughter. Clinging to the handlebars of Jimmy's bike as they sped downhill from the schoolhouse. "Oh, the padded crossbar is the man! The McKevitts always like to travel in style!" he cried as he dared to nuzzle into her windswept hair . . .

"Holy God! I'll be flying home every weekend if I get a welcome like this," Jimmy said, disengaging from her but still resting his hands on her shoulders.

"Flying?"

"Sure thing. I came on the flying boat to Foynes in Limerick. It was a trial flight, but I've got connections. Man! What an experience! Thirteen hours flying. A bit rocky at times, I tell you. I'll be glad to go back on the liner!"

"Thirteen hours? You came to Ireland – "

"Can you believe it? Took me all of that to get from here to Queenstown last time."

"God in heaven! Look at me here, gabbin' away and never even asked you to have a drink or – "

"I'll have a cup of strong tea and a slice of your soda bread – that's if you still make it?"

She led the way into the kitchen.

"Indeed I do," she laughed, "even if it's only for myself . . . "

"I – I heard about Podgy. Poor ould Podgy. I'm sorry."

"Aye." Bridie rattled a poker through the bars of the range. "That's the way of it." She busied herself tidying the kitchen.

195

Jimmy loosened the belt of his coat.

"Would you ever just sit down, Bridie, so I can talk to you?"

"I'm sorry. The place is a bit of mess."

"The place is grand – and it's you I came to see."

"Bold as ever," she laughed, as she seated herself by the range. "Now, what really brought you home?"

"That's your trouble, Bridie McKevitt. You never did believe me!" He stared solemnly at her before breaking into a smile. "The old boy's dying – "

"I heard he wasn't – "

"God knows I only came because the old lady wired me. Do you know what his first words to me were?" Jimmy's tone became suddenly bitter. His accent reverted to his native Monaghan. "'Will ye g'out,' says he, 'and give Owney a hand with the milkin'.' No such thing as 'You're welcome home,' or 'How is America.' No sir. It's 'will ye g'out and give Owney a hand.' Jeez!"

"His mind's a bit scattered, Jimmy." She felt sorry for him as he searched his pockets for cigarettes. The kettle began to sing on the hob.

"He was always a bit scattered where I was concerned, the old bugger. Good old Jimmy. Lend a hand. I'll probably be sent out thinning turnips tomorrow –"

"At least you're dressed for that!" Bridie decided to try to humour his bitterness. It worked. He exploded into laughter. The old Jimmy.

"Godammit, you're right. Can't you just see me on my knees in a drill of turnips in this get-up?"

Bridie took the soda bread from an old biscuit tin and began to slice it.

"America has been good to you, Jimmy."

"Eventually. It was rough for a long time. Bloody rough. But I made a few investments – and they paid off. So maybe I should thank the old boy for shunting me off to the States. The only thing is I should have taken you with me."

Bridie blushed as she put the tea and soda bread before him.

"Anyway. Enough about me. How are you? And Brian? Where's Brian?"

"Brian's away at boarding school. He got a scholarship."

"He was always a bright lad. Can you manage this place on your own?"

"Me and Hanratty get by."

"Is he still here?"

"I'd be lost without him."

"Jeez, Bridie. This soda bread is out of this world. No one ever came near you for soda bread." He hastily buttered another slice. "So things are OK for you?"

"Why do you ask?"

"Because I care."

Bridie turned away quickly and took a few sods of turf from the cardboard box by the range. She fumbled with the lever in an attempt to lift the plate from the top of the range.

"Here. Let me do that!" Jimmy sprang to her and took the lever from her trembling hand and, as he prised the plate open, he noticed her tears glistening in the reflected firelight. He dropped the turf into the fire and replaced the plate.

"Now, sit down there," he said very softly, "and tell me what's troubling you."

"It's – nothing."

"Just the sight of me reduces you to tears!"

"No." No. No. No. Just the need to confide in someone. To share the worry that kept her awake, night after night.

"For Christ's sake, Bridie, what is wrong?"

She wanted to speak but couldn't. She stood up, brushed past Jimmy and made her way out to the bar. She retrieved the letter from behind the trophy, paused and was about to replace it when Jimmy's arm reached over her shoulder to take it from her. They returned to the kitchen as Jimmy read the letter with growing disbelief.

"For Christ's sake, who is this guy?"

Bridie shook her head.

"I only met him at the funeral. I'm sorry. I shouldn't have shown – "

"Of course you should. These guys need sorting out." He folded the letter and slid it into an inside pocket.

"What are you going to do?"

"You're talking to James McCabe, law student, ma'am," he drawled. "Five years at night school. As for that McGribben guy, he needs a good kick up the rear end. Just leave this to me."

Bridie felt almost giddy with a tremendous sense of relief.

"Here, have another slice of soda bread." She knew as she said it that it sounded silly.

"Mmm! Don't mind if I do!" He ate with relish. "You know you're just going to have to give me the recipe for this." He chased the remaining crumbs around the plate. "I want to bring it back to Mary Kate."

"Oh." Bridie felt the colour draining from her face. He noticed her reaction instantly.

"You didn't know? Godammit, Bridie, I thought, I mean – did they not tell you?"

"I suppose I've been a bit confined here over the last while." Her voice was little more than a whisper. "Podgy was a lot of – took all of my time."

"Dammit!" Jimmy reached for another cigarette, then spoke in a measured voice.

"Mary Kate is – was – Mary Kate Boyle. Her father came from Donegal. We met in New York. She became – Mrs McCabe three years ago." He drew long and hard on the cigarette. "She's a nice girl. You'd like her." He exhaled a cloud of smoke. "We have a little girl – Anne – she'll be two in January."

The silence that followed was broken only by the sound of the turf collapsing into the heart of the fire.

"Would you like more tea?"

"No, thanks. I must be away." Jimmy consulted an expensive-looking wristwatch. "I'll probably be needed to dig a few drills of spuds before dinner."

Bridie smiled weakly at him. He patted the pocket where he had put Bridie's letter.

"You leave this to me. I'll be in touch."

She did not see Jimmy again until his father's wake a week later, when she called to pay her respects. A rosary was in progress when she entered the house. Hunched and kneeling figures were silhouetted in the flickering candlelight, making hushed responses to Minnie McCabe's Hail Marys. When the prayers were finished, the neighbours shuffled out to the kitchen and parlour where food and drink were offered. Bridie shook Minnie's hand warmly.

"I'm sorry, Minnie."

"I know that, Bridie. Och, he was a decent man, whatever. 'Tis not easy losing your life's partner, but sure, you know all about that."

A voice interrupted them.

"Bridie. Good of you to come." She shook Jimmy's hand awkwardly. He stood before her, holding a bottle of whiskey in one hand. "Will you take a drop?"

"A drop of tea will do me grand. I'm – sorry."

He cupped her elbow in his hand and guided her to a quiet corner of the room.

"Ah sure, the old bugger. He had a hard death. We're all sorry. I'm just sorry he didn't give me any credit for anything, years ago. He thought I had no sense – now where did he get that idea?"

They both smiled. Jimmy looked around anxiously before he continued.

"Anyway, about your own bit of bother. That's all settled up. You'll have no more of them letters – "

"But how – ?"

Jimmy exaggerated his American accent.

"Ah reckon ah just blinded them with mah superior legal knowledge. Quoted them several cases to prove mah point."

"Oh Jimmy. I can't believe this. I can't! I can't." She searched frantically for a handkerchief.

"For God's sake, Bridie. They'll all think you're crying for the old boy!" He put his arm around her shoulders. "It was just a favour I did – for a friend. Maybe to make up – to make up for favours that weren't done before."

She looked at him with a puzzled expression. It was the nearest Jimmy McCabe had ever come to saying sorry. Bridie dabbed her eyes quickly.

"I – I have to go. I told Hanratty I'd not be long. I brought a bit of bacon" – she gestured towards a box under the table – "and a couple of loaves of soda bread." His eyes sparkled at the mention of soda bread.

"Well, they'll be smuggled onto the liner for a start."

Bridie buttoned her coat quickly and made for the door while the path was clear. Before she put on the second glove, Jimmy took her hand in both of his.

"I wish you happiness, Bridie."

"Thanks, Jimmy." She hesitated for a moment and thought of kissing him but Owen Brady and his wife arrived at that moment.

"There y'are, Jimmy," Owen said as he handed him a parcel. "The tricolour for your father's coffin. We mustn't forget those who fought the fight when it was needed. How're you, Bridie?"

Bridie nodded towards the Bradys and set about extricating her bicycle from the heap that was stacked against the wall. Jimmy came to her aid.

"Thanks again – for everything," she whispered.

"Mah pleasure, honey bunch," he drawled.

She feigned a scowl as she rode into the gathering gloom. He stood there, framed in the doorway, waving until she disappeared into the night. The November air pinched her face as she sped homewards but she knew that the tears streaming down her face had nothing at all to do with a winter wind.

The letter arrived a few days later. The handwriting was all too familiar. She couldn't wait to get rid of Owen Brady who was in a particularly reflective mood about the "struggle for freedom" and "the debt owed to the Charlie McCabes of this world." She tore the letter open before he had closed the pub door.

Dear Bridie,

A hurried note before I set off for Queenstown. I need to get back to the business and there's no point in hanging around at home any more. I was always in the way there anyway. If only I had sense in my youth! I'd probably be the proud owner of thirty acres of rushes now and be beating the women away from my door! Life is full of "if onlys", Bridie, but I suppose we make the best use of the cards we are dealt.

It was great to see you again. And to be able to help you out. If ever you need help (or if you ever make it to the US of A!) my address is on the other side of this page. I am writing this note rather than call in to see you, because goodbyes get harder with the years. I take a part of you back with me – and I don't just mean the soda bread.

May you have happiness, health and luck always, Bridie. No one deserves it more.

Yours ever,

Jimmy.

She folded the note carefully. That evening she slipped it between the leaves of *The Works of Robert Burns*, at page 120 where the book naturally fell open. She didn't want to read that poem again but three lines jumped out at her:

"Now thou's left thy lass for ay –
I maun see thee never, Jamie
I'll see thee never."

Two days later Hughie called in to the shop with a list from Josie.
"Well, what did you think of the Yank?"
"The Yank?"
"Ah come on, Bridie. You know I mean – "
"He seems to be doing all right."
"You might sing that! There's a fellow landed on his feet all right."
"What's this at the bottom?" She handed the list back to her brother, who examined it closely.
"Four ounces. No – no – what am I saying. Forty cigs for James," he laughed.
"He's a terror at those cigarettes. I suppose there'd be no cobbling done if he didn't have a smoke."
"Not a stitch." She handed him the cigarettes.
"You were saying – about Jimmy?"
"Oh aye. Landed on the two feet. Picked the right woman – "
"Fourteen and tenpence. Fourteen shillings even," she added coldly. He tendered a pound note and continued.
"Her old man owned a few restaurants. The bould Jimmy took them over and expanded the business. 'Diners' they call them over there. Has a whole string of them – all called 'Jimmy Mac's.' Would you credit it?"
"He told you this?"
"Aye – the night of the funeral. The drink was flowin'! Oh listen, you'd better give me a few penny toffees for the weans or I'll be in trouble."
Bridie scooped a handful of penny bars out of the box. "There's one there for yourself as well." Her sarcasm was lost on him. "They're on the house."

201

As Hughie left, Bridie felt her cheeks burning. Was it with rage or frustration? She couldn't tell. Hughie had no sooner closed the front door when Hanratty staggered in the back door with a crate of bottles.

"God in heaven! Is it possible to get any peace in this place?" Bridie muttered as she slammed down the counter flap and stormed into the kitchen.

CHAPTER TWENTY-SEVEN

THE HUBBUB DIED AWAY QUICKLY IN THE STUDY HALL. SUNDAY MORNING. Eleven o'clock. It was letter writing time. Father Laurence paced to and fro on the dais at the top of the hall, peering over his glasses in search of the few who continued to make noise.

"Messrs McCormack and Brady!" he bellowed at two boys who were engaged in heated conversation in the middle of the hall. "We would all prefer if you began writing your letters this Sunday rather than next! What seems to be the problem?"

"I have no pen, Father. He fec" – sniggers all around – "he stole it on me," McCormack called back, nodding towards his rival.

"I did not," Brady countered. "I never even seen it."

"I'm tired of these petty squabbles," Father Laurence sighed. "And equally tired of the strangulation of the English language. Can anyone lend this child a pen?" A hand went up behind McCormack.

"Good! Now could we all get on quietly with the business of the morning?"

Silence was soon restored, save for the occasional squeaking of a desklid or the rustle of paper. Father Laurence began reading his breviary.

Brian chewed on the top of his pen before beginning to write.

25th October 1937

Dear Hannah,

Here I am in boarding school. St Felim's is all right so far. I suppose I am used to being away from home so it was easier for me

than it was for some of the other lads. I got ducked in the fountain by
the seniors but they do that to all the first years. I play football in the
junior leagues but we haven't won a match yet! I am in the choir too.
Father Vincent says I have a very good voice. We will be putting on a
concert at Christmas.

I hope you are getting on well at your school And I hope you will
write and tell me about it. I hope Hazel is not teasing you too much.
I must go now because I have to write to Aunt Bridie too. I will send
this to Sloe Hill and hope they will send it on to you.

Your fond brother,

Brian.

PS Fifty-seven days to go to the holidays. Roll on Christmas.

He folded the letter and slid it into an envelope. He addressed the
envelope and left it unsealed on the corner of his desk. At the end of
letter writing a prefect would collect all the letters. One of the priests
would check the letters before they were posted. Father Laurence had
stormed into Brian's class one Monday morning and biffed Slim
Monaghan for writing to Mr Joe Louis, New York, America, for tips on
how to become a boxing champion.

"Shame on you, Monaghan," Father Laurence scolded the red-
faced boy. "Showing an interest in such a savage sport and then
writing to a black man about it. A black man," he jeered, as he tore
the letter up into small pieces, showering Slim with the confetti.

"Pick those up!" he snapped, "and write an essay for me by
Thursday – "

"Ah Father – "

" – on the topic 'Boxing Belongs in the Jungle'."

"Ah Father – "

"At least two pages."

Father Laurence never entertained the possibility that someone
else had written the letter to Joe Louis in order to get Slim into
trouble. Slim was grossly overweight and had not the slightest interest
in any athletic pursuit. It was probably Lacy who wrote the letter,
Brian thought. Lacy liked making trouble. He was older than the rest
of the class. It was said that he had already been expelled from
another school. Brian was afraid of Lacy and kept out of his way.
Although he was older than the others, he was smaller than most of

them, being of squat build. His nickname was Porky – someone said it was because he had eyes like a pig's – but very few would use it to his face.

Brian was halfway through his letter to Aunt Bridie when Father Colman, the President, strode through the study hall and ascended the dais. Father Laurence left the hall.

"Pens away, boys. Five points!" Father Colman believed in order and economy of words.

"Point one – work. This is not a holiday camp. You are here to work. Your parents have made great sacrifices. Your teachers have made great efforts. You must make great efforts too. Christmas exams in six weeks. Get down to work!

"Point two – smoking. Smoking is forbidden in this school. That is the rule. The penalties for breaking the rule are severe. And yet the senior toilets are like a Red Indian reservation each lunch-time with smoke signals rising in clouds. Red Indians – be warned! Cut it out!

"Point three – letters. Letter writing to the opposite sex is not permitted. Three such letters were intercepted in the past week. Do not attempt to use day boys as postmen. You are endangering your future and theirs in this college.

Point four – pilfering. There's a lot of it going on. Be careful with your possessions. Be vigilant and report any suspicions you may have. Your desks and lockers may be inspected without warning.

"Point five – punctuality. There are certain laggards amongst us who dawdle into morning Mass, arrive late for class, for showers, for choir practice – for everything but meals. They know who they are. More importantly – I know who they are. Cut it out or else – you may be first down to the refectory but you will find there is no food for you.

"That is all. Continue writing."

Choir practice was held after tea on Sunday evening. For Brian it was a welcome break from study. Father Vincent, in his quiet, unruffled way, took the firsts who included Brian, through the *Ave Maria*, then the seconds and finally the thirds.

"Good!" He seated himself at the piano. "Now, boys. We'll put it all together."

They sang uncertainly at first but a second, then third, rendering gave them confidence.

"Grand, boys, grand! A little more polish and we're there. Now for a wee bit of Scotland. Have any of ye heard of Robbie Burns?"

Brian's hand went up involuntarily.

"Good man, Brian. What do you know of him?"

"He wrote a whole lot of poems and songs. My Aunt Bridie has a book of them at home."

"Indeed he did and we're going to learn one of them tonight. It's called 'My love is like a red, red rose'."

When practice was over, Father Vincent called out four names – Paul Kelly, Conor Maguire, Brian Johnston and Art McBride " – stay back please." The four looked at each other in puzzlement.

"Don't worry, lads," Father Vincent laughed. "You did nothing wrong! In fact you did everything right. I just want you to come up to my room for tea and music – a little treat! At least I hope you'll consider it a treat!"

They followed him down a dark corridor into a spacious room.

"Now, sit yourselves down and we'll have a little soirée!" He filled a kettle and put it to boil on a stove.

"What's a soirée, Father?" Conor asked.

"A little chat and a little music. I just want to introduce you to some music, boys. There's more to music than George Formby, you know!"

The boys laughed nervously as he moved to a piano in the corner of the room.

"Just relax, boys, and listen to this."

His long slender fingers coaxed a gentle tune from the piano. He closed his eyes as he played. Conor Maguire, seated behind the priest, mimicked his playing but Brian felt himself drawn to the music.

"This is a nocturne, a piece of night music written a hundred years ago by a man called Frédéric Chopin," Father Vincent whispered. Brian sat back in his chair. The flames in the stove seemed to dance in time with the music. He too closed his eyes.

When the music ended, Father Vincent made tea and produced a packet of biscuits from a cupboard. While they ate, Father Vincent told them a little about Chopin's life including his love for George Sand.

Conor sniggered at the mention of the name.

"It's not what you think, Conor. George Sand was a woman, a novelist. That was her pen name." Father Vincent stirred his tea slowly.

When they had finished tea, he resumed playing.

"Another Chopin piece. A little *étude*." In the middle of the piece, he suddenly said, very quietly, "I suggest you leave now, gentlemen. You mustn't be late for night prayers."

"Thank you, Father," the boys replied in equally hushed tones. He nodded and continued to play as they left. The other three raced down the dark corridor, giggling as they went. Brian reluctantly closed the door of Father Vincent's room behind him.

Mr Carroll, the Latin teacher, burst into classroom 1A, rubbing his hands fiercely. His round pink face peeped out over an extremely long scarf that was wound several times around his neck. He continued to rub his hands while he solemnly announced to the class in a sonorous tone, *"Hiems adest, pueri!"*

There were a few sniggers, then silence as he stood facing them, expectant.

"I am awaiting a translation, gentlemen. *Hiems adest, pueri!*"

He turned to the giggling red-headed boy by the window.

"Perhaps Mr Corley will share with us his vast knowledge of the classics?"

Corley looked up sheepishly.

"The boy made a hames of his sums, sir," he spluttered.

Mr Carroll winced. The pink face sank deeper into the scarf.

"O tempora, o mores," he sighed. "Is there not one among you, the cream of Ulster's intelligence, who will translate for me the simple statement – *Hiems adest, pueri?*"

He rubbed his hands again.

Brian raised his hand nervously. Mr Carroll's face brightened.

"Johnston, *salvator meus!*" he said hopefully.

"I think it means 'Winter is here, boys'."

Mr Carroll bowed graciously to Brian.

"*Supremus inter pares,* Johnston!"

Brian shrugged his shoulders with embarrassment. He had partly guessed the answer, but from his first encounter with Latin he had taken to the subject. He loved its strangeness and its construction and he enjoyed finding connections between Latin and English words.

"Page 125, boys. *Viva voce*," Mr Carroll sighed. "Indeed – *hiems adest sed non habemus calorem*. I am tempted, like Caesar, to set out into hither Gaul."

When Father Laurence left the dormitory to collect the mail, the noise level rose immediately. Beds were tossed, pillow-fights began and when someone produced a football a general mêlée ensued. The ball bounced under a bed and suddenly Lacy's strident voice was heard above the mayhem.

"Ah jeez, Dandelion, the stink!"

The boy standing beside the bed turned away in shame and pretended to look for something in his locker. Lacy whipped back the bedclothes to reveal the dark brown rings of stain on the sheet.

"Ye little pissybed! Why don't you tie a knot in your mickey?" The noise died away and Lacy, realising he was the centre of attention, held his nose in mock disgust.

Three beds up from Dandelion, Brian had a fleeting image of Johnny Caraher kneeling at his bed in Drumcran, draped in stained sheets.

"It's not his fault," he said nervously.

"Who asked you?" Lacy snapped.

"No one. I'm just telling you. It's not his fault. Why don't you leave him alone?"

"Why don't you just mind your own business, or go up and have tea with Vincent, little Nancy-boy?"

Brian gripped the blanket fiercely. Lacy's remark stung him. He felt his face grow hotter as attention was turned on him.

"Nancy-boys and pissybeds," Lacy taunted them.

A shout from the door warned them of Father Laurence's return. There was a scuttling to and fro and a hasty rearrangement of beds. When Father Laurence entered, order had been fully restored. He marched briskly down the passageway tossing letters to left and right on occasional beds.

"Lights out in five minutes, gentlemen – and that includes you, Lacy," he announced as he disappeared into his room, a walled-off section in the corner of the dormitory.

A letter landed on Brian's counterpane. The neat, rounded

handwriting with each line of the address underlined told him it was from his sister. He snatched at the letter. The envelope had already been slit open.

Lambeth Gardens
Belfast
Nov 12th 1937

Dear Brian,

I got your letter from Aunt Rose. She came up to Belfast for Hallowe'en. I am glad you like St Felim's. I don't think I would like all them priests watching me all the time. I am in the choir too in Miss Dagg's. The senior girls are called Daggers. We are called Blades. The school is all right but we get lots of homework. Hazel is always annoying me looking for answers and copying what I do. She meets a boy called Norman every evening after school. I have to wait around for her so that we both arrive home together.

I am looking forward to Christmas. I have to go now because Hazel keeps trying to read this. She thinks I am writing to a boyfriend. I will keep her guessing.

Your fond sister,

Hannah.

PS Uncle Winston says there's going to be a big war. What do you think of that?

CHAPTER TWENTY-EIGHT

"COME ON!" HUGHIE CALLED. "YOU'RE LIKE A CLATTER OF OULD WOMEN chattering away there. There's rain on the way and hay to be saved." The three boys waved back to acknowledge his words.

"We'd better hurry," Peter Joseph said from atop the half-built haycock. "If we don't get this field cocked this evening, Daddy will kill us!"

"Ah, let him!" Micky McMahon barked as he forked a bundle of hay up to Peter Joseph.

"We're doing our best," Brian said, looking glumly from the remaining hay-rows to the darkening sky. "It's hard enough work."

"Hard work?" Micky jibed. "You wouldn't know anything about hard work and you living the gentleman's life most of the year – "

"Gentleman's life, my eye!" Brian laughed.

"Come on!" Peter Joseph called. "Daddy's watching us again."

Micky ignored him.

"You'd never last it if you were like me – out in all weathers all year round." He jabbed his fork into the stubbled earth and leaned on it. "I hate it. I wish to God I could get a job."

"You could try for postman," Brian said. "Uncle James says Owen Brady was lifted for being in the IRA. He's away in prison in the Curragh."

"Naw! I'd still have to come around here. Every time I'd come by I'd have to dig spuds for my mother or feed the calves for my father! I was thinking of joining the British Army – "

He pointed the fork handle at Peter Joseph. "Pitaow! Gotcha!" He saluted Brian. "I got that Jerry sniper, sir."

"You're only fourteen, Micky," Brian laughed.

"Well I'll join the Jerries, then. I'll write to Hitler. I hear he's looking for spies. I seen a great picture in Monaghan – "

"Oy!" Hughie shook his fist threateningly at them from the other side of the field. The first spatters of rain slapped into Brian's face.

Uncle James's loft was a welcome haven from the rain and Hughie's ire. Brian loved its cosiness, its seeming disorder and above all the reassuring smell of leather and wax.

"Uncle Hughie's in a bad mood because we didn't get the hay finished."

"He'll get over it." James lit a cigarette and began searching in a cupboard underneath his workbench. "I have something here for you – if I can find it." He took out a small box and from the box he took an eggcup.

"You fixed it!" Brian cried with delight.

"Well don't look too closely," James laughed. "There's still one little chip missing, but it's well glued together. Don't let it fall again . . . "

Lacy slammed the door of his locker shut. "All right," he roared. "Which one of you feckers took my toffee?"

There was an uneasy silence in the dormitory.

"I had a whole slab of toffee in here. It didn't walk out on its own." Nobody spoke.

"All right. If that's the way you want it, I'm entitled to search for my property." He looked around slowly. "I'll start with you, Johnston." He marched up to Brian's locker.

"You're not going to search – " Lacy thumped him in the chest, knocking him back across his bed.

"Who's going to stop me?" Lacy flung the locker door open and began rooting out the contents.

"I'm only looking for what's my own. A slab of – Well, well, well. Look what we have here!"

He held an eggcup up for all to see.

"Give me that!" Another thump sent Brian sprawling again.

"Ah look!" Lacy sneered. "Mammy gave him his own little eggcup!

211

With little pink roses all over it. His own little pinky-eggycup. Oh dear!"

Brian shut his eyes tightly, wishing he could shut out the sound also.

"Oh dear! I've dropped the little eggy – "

Blind with rage, Brian hurtled his body at Lacy, catching him off-balance. The two fell across the next bed. He pummelled Lacy's face, screaming, "I'll kill you! I'll kill you!"

The other boys gathered around quickly.

"Row! Row! Row!" they chanted.

Lacy twisted and turned in a desperate attempt to avoid Brian's fists. He grabbed Brian's hair and pulled with all his strength. The two toppled over between the next two beds. Now Lacy was on top. His fist smashed into Brian's jaw – once, twice –

"What's going on here? Get out of there, you pair of savages!"

Father Laurence's angry voice sent the other boys scurrying back to their beds. He grabbed Lacy by the collar and hauled him upright.

"Trust you to be involved, Lacy!" He slapped him viciously across the face. Father Laurence's red face registered genuine surprise when he saw Brian struggle to a sitting position on the floor.

"Johnston! I wouldn't have thought you would succumb to the laws of the jungle! Well! I'm waiting for an explanation for this . . . "

Lacy threw a dark look at Brian.

"It was just a – a bit of horseplay, Father," he muttered.

"Indeed! You demean that noble animal, Lacy. Well, Johnston?"

Brian sat dazed on the bed. His jaw stung terribly. He said nothing.

"Right. The two of you can horseplay up to the President's office in the morning. Into bed, the lot of you!" He strode to the door and switched off the lights. "Savages!" he muttered as he marched into his room.

Brian eased his aching body into his bed. He began to shake all over and could only barely discern the figure that suddenly appeared at his bed. For a terrifying moment he thought it was Lacy until the figure handed him a crumpled page of a copybook.

"That's all I could find!" a voice whispered. It was Dandelion. The page fell open. In the faint moonlight Brian could see and feel one, two, three, four pieces of jagged china. He slid them under his pillow, drew the counterpane over his head and cried quietly into the pillow for a long time . . .

"I said – do you remember this? God, you're a million miles away, Brian." James was holding up a small hammer. "The wee hammer! When you were a wee fellow you used to come up here every day and ask for the wee hammer. 'Where's the wee hammer, Uncle James?' you'd say. And you'd hammer away, happy as the day was long."

"I was only four or five then," Brian said, slightly embarrassed.

"Aye, you're a big man of fourteen now."

"I'll be fifteen in two weeks' time. Can I put on the wireless, Uncle James?"

"More of that ould music, I suppose?"

Brian tweaked the tuning knob through a range of stations. The names on the dial had always fascinated him – Hilversum, Strasbourg, Monte Carlo . . . Suddenly he stopped on recognising a familiar tune.

"That's Beethoven – *Moonlight Sonata*. Father Vincent played that for us!" he cried excitedly.

"Give me Bing Crosby any day," James said through a mouthful of tacks. Brian was oblivious of the rhythmic tapping on the last as he sat watching the rivulets of rain careering down the windowpane . . .

"Well, what have you pair of miscreants to say for yourselves?" Father Colman snapped as he ushered Brian and Lacy into his office. Lacy began muttering.

"Speak up, boy!"

"It was a bit of horseplay, Father."

"Looking at the damages incurred, highly unlikely! Johnston?"

"Horseplay, Father." His jaw hurt when he spoke.

"I see." He reached to the top of a bookshelf and took down a long cane. "Well, horses need to be whipped sometimes, don't they?"

The boys remained silent.

"Hands!"

The cane swished down six times on each hand. When it hit his palm, Brian could bear the stinging pain but when it caught his fingertips his eyes watered as the searing fire burned into the flesh. By the time the last stroke came there was total numbness in both hands. Lacy gave a little yelp after each stroke and desperately squeezed his hand under his armpit.

"Come on, Lacy. I have more important things to be doing." One more thwack and it was over.

"Out of my sight, both of you. And if I see you up here again, Lacy, pack your bags first."

Lacy ran back down the corridor to the study hall. Brian went through the side door into the college grounds. He made straight for the fountain and held his swollen hands under the cascading water. Cooling, healing water . . .

Bridie put the newspaper down on the kitchen table. "God preserve us all," she sighed. "Them three poor women in Wexford. Killed by a German bomb. Are we all going to be killed in our beds?"

Brian took up the paper.

"I heard Mr Churchill on Uncle James's wireless," he said. "He says they'll beat the Germans."

"Well he'd want to hurry up! Now we'd better get your trunk packed for school. Where did this summer go at all?"

She busied herself sorting sheets and pillow-cases. As she folded them and packed them neatly in the bottom of the trunk, she observed Brian seated at the table. When he furrowed his brow in reading the newspaper she could see Sarah poring over her English book and chewing the top of her pencil . . .

"Bridie, what's e - t - i - q - u - e - t - t - e?"

"I don't know."

"It's a kind of fancy manners," Mother said from the scullery.

Father turned the bellows wheel by the fire. "Like blowin' your nose with a handkerchief instead of your thumb," he said.

"Eti-quette! Eti-quette!" Sarah savoured the word. "Does Ettie Quinn have Eti-quette?" she giggled.

"Wouldn't know it if it jumped up and bit her," Father grunted.

Man in the moon, man in the moon
Why did the dish run away with the spoon?

On their first evening in the senior dormitory, there was a buzz of excitement among the boys.

"Did ye hear?" Corley said. "The Jerries are after bombing London!"

"I hope they come over there and bomb this feckin' place," Conor Maguire chuckled. "Can't you imagine it! Larry would be running around shakin' his fist at them and shouting 'Savages! Savages!' And The Coalman would bring us all up to the handball alley. 'Two points, boys. Point one. We have no school left. Point two. Yis can all feck off home!'"

There was an outburst of raucous laughter and cheering. When it died down, Art McGrane's quiet voice intervened.

"Lacy's not coming back."

"How do you know?"

"I met him in Monaghan. He's working in a butcher's shop."

"I hope no one complains about the meat, or they'll be carved up!"

Brian gave a sigh of relief and glanced across at Tommy Connolly who gave him the thumbs up sign. Perhaps he could now cast off the name Dandelion, Brian thought.

"Point three – there will be no choir practice until further notice."

The rest of Father Colman's words were lost on Brian. He looked across at Art who shrugged his shoulders. Was Father Vincent ill, he wondered. On Monday evening he was delighted to discover that this was not so when he noticed the priest, deep in thought, walking slowly down the avenue. On an impulse Brian ran towards him.

"Father Vincent – "

"I trust you realise, Mr Johnston, that if you set foot on this avenue, you are out of bounds!"

Brian stopped abruptly.

"But – the soirée?"

"You may tell your friends to report to my room at nine o'clock as usual. I will ask Father Laurence to excuse you from study. Now run along."

Brian was accustomed to Father Vincent's varying moods but there was a strangeness about him this evening that was perplexing.

"Just in case you think all composers are dead, we'll begin this evening with one that is very much alive. His name is Sergei

Rachmaninov, a Russian. This is one of his variations on a theme by Paganini. Did you get that, Mr Maguire?"

"Yes, Father," Conor answered doubtfully.

"Just enjoy the warmth of it." He played with a delicacy of touch that transfixed Brian.

"That was lovely, Father," he whispered as the music ended. Father Vincent's eyes remained closed. His hands rested on his knees.

"I'm glad you approve," he said after a long silence. "It was only written a few years ago. I was lucky to get the sheet music from a friend in London. I've been practising it especially for you, for this, our final soirée."

The boys looked at each other in surprise.

"Are you leaving us, Father?" Conor asked.

He rose to prepare the tea. "Yes. I am being sent to England."

"God, you'd want to watch out for the bombs over there, Father," Conor said with genuine concern.

"At this stage I am used to bombshells, gentlemen." He set the china cups before them. "It seems somebody suggested to the bishop that I was a danger to young men – "

"How do you mean Father?"

"I suggest you ask the bishop that – or his anonymous correspondents. Anyway, parish work in the British midlands beckons."

He poured the tea carefully.

"To mark the occasion I have managed to procure a cake, so eat and enjoy, gentlemen."

They wolfed down the Madeira cake that had been neatly sliced into four portions.

"Will you not have some yourself, Father?" Brian asked.

He shook his head and seated himself at the piano again.

"This is my food. I will miss our little soirées, gentlemen. I enjoyed your company. Most of all I will miss my beloved Bechstein." His slender fingers caressed the keys lovingly. "To end, a piece from the master – Mozart. The Andante from his C Major Concerto."

He seemed to surrender himself totally to the piano. The fragile cadences expressed the pain that was evident in his face. Brian bit his lip. Once he thought he saw a tear glistening on the priest's eye. The music totally overpowered him. When eventually Father Vincent said in his usual quiet voice, "Time for you to go now, gentlemen. Thank you and goodbye," Brian could not move. The others said a muted "Thank you" and left. For once there was no mad rush down the corridor.

"Father, do you have to – "

"Please go, Brian."

He fought back the tears.

"Just think of me sometimes – as I will think of you."

"Yes, Father."

"Now go or you'll be late – "

"I don't want to go, Father."

"Go! Now!" he snapped.

Brian left quietly and walked down the corridor in a daze, the haunting strains of Mozart echoing behind him. He could not face night prayers. He went down the back stairs and made his way to the dormitory where he flung himself on his bed and sobbed loudly.

CHAPTER TWENTY-NINE

THE MEN AT THE BACK OF CULLYBOE CHURCH WERE GROWING INCREASINGLY restless. It was bad enough having the long gospel on Easter Sunday morning but on top of that Canon McKenna's sermon had gone on for a full twenty minutes. Shuffling feet and a barrage of coughing had not deterred him.

"And so we rejoice on this Easter morning, my dear people. On this great feast of the resurrection of Our Lord, we are filled with hope and promise. We renew the vows of our baptism. We renounce the world, the flesh and the devil – all three of which sadly assail us all the more in these times. I call on the young people especially to stand up for your faith, for the old values. I passed through Monaghan last week and saw a cinema notice of a forthcoming presentation. I cannot even bring myself to repeat the title of the film in the house of God. Suffice to say the very title promoted evil and wanton ways and I hereby enjoin all of you to shun it, boycott it, have nothing to do with it.

"I'm sure you don't need reminding that we have another reason for rejoicing this Eastertime. Twenty-five years ago, brave and noble men were put to death by the Saxon foe for their part in the Easter Rising of 1916 . . . "

Petesy Marry, leaning against the statue of St Anthony, wrung his rolled-up cap in frustration.

"Holy Christ," he muttered. "If he goes on much longer there'll be a bloody Easter Rising in 1941 – in Cullyboe . . . "

218

" . . . so we celebrate the memory of those brave men who were prepared to lay down their lives for a cause they believed in. Greater love than this no man hath . . . The foe may have changed but it is all about us – in the dance-hall, the cinema, the books – and it is more insidious than ever. Stand up, my dear people, and fight it!" He paused and waited for the coughing to subside.

"The Easter offerings are due and the list will be read out next Sunday . . . "

When they returned from Mass in the pony and trap, Bridie and Brian only stopped at the house long enough to collect the box of food and sweets Bridie had prepared for their visit to Drominarney. Josie had invited them for Easter dinner. Brian urged on the pony as they sped along the road under trees that were just coming into leaf. He relished the prospect of Josie's cooking, playing in the hill field with his cousins and listening to Uncle James's wireless. As they drove into the haggard there was an unreal quiet about the place. Something was wrong.

Josie suddenly dashed out from the scullery, wiping her face with her apron. She was crying loudly.

"Oh Bridie – "

"What is it? What's wrong?"

"It's the cattle," she sobbed. "The foot and mouth. They all have to be – slaughtered."

Brian leaped from the trap and ran to the cowshed. Inside the younger children were sobbing and sniffing as they stroked and petted the four cows. He made straight for the roan cow. How often had he milked her, foddered her, waited with her while she calved? And now . . .

She didn't even look sick. As he squeezed her ear, her big docile eyes came around to face him. He could feel her warm breath on his face. He buried his face in her neck and felt his tears run down her silken hide.

Hughie and Peter Joseph entered the cowshed.

"Get out! Get out of here! The lot of yis!" Hughie roared at the children. They cried even more loudly, frightened by their father's unaccustomed anger.

"I'll take them off for a wee drive in the trap," Bridie said.

"You could take them down to my mother," Josie suggested. They left reluctantly.

Brian had never seen Hughie cry before.

"What are we going to do, Brian?" he asked, between angry sighs. "All of them – gone."

"H-how?" Brian asked.

"I don't know. It's all over the country. I thought we'd avoided it. I thought – "

Another trembling sigh. He put his hand on the roan cow's rump.

"Best milker I ever had. Best calver. Reared her from a calf myself . . . It's not fair," he sobbed. "It's not bloody fair."

On the following morning Josie arrived at Bridie's shop with the three younger children.

"The sergeant is coming this morning," she explained. "I don't want them to be around for that. I don't want to be around myself either."

Bridie ushered them in and, having shared out toffee bars among them, she gave the children the freedom of the house.

"I'd like to go up to Drominarney," Brian said, when the children went out to watch Hanratty at work in the bottling shed.

"Are you sure?" Bridie asked. "It's not going to be nice up there."

"I'd rather be there."

Sergeant Carney arrived shortly after noon with an ominous-looking bundle strapped to the carrier of his bicycle.

"They'll not suffer, Hughie," he said as he unwrapped the humane killer from the sacking that covered it. To Brian's eye it looked a crude and uncertain weapon – a heavy mallet and a strangely shaped object which the sergeant was already loading with a cartridge.

"You and me can do this job alone, Hughie." He nodded towards Brian and Peter Joseph.

"Go on up to James, you two," Hughie said as he fiddled nervously with a length of rope.

"I want to – " Peter Joseph began.

"Go on – now!" Hughie snapped.

James sat, cigarette in mouth, crouched over a bridle he was attempting to stitch. The entry of the boys told him that the moment they all dreaded was finally at hand. He pushed away the bridle, reached into his pocket and withdrew a rosary beads.

"I think, boys," he said quietly, "at a time like this all we can do is pray. We'll try a decade or two to help us all – and that the animals won't suffer."

They had only got halfway through the first decade when the first dull report came from below. The boys faltered in their response until James took up the response himself. As they struggled on, a new sound assailed their ears – the terrified cry of animals who sense death. A second report was followed by bellowing that none of the three had ever heard before.

"Christ, this is no use," James cried. "Find us some music on the wireless, Brian."

He lit another cigarette with difficulty as his hand shook terribly. It seemed as if the wireless would never warm up. Peter Joseph clamped his hands over his ears and stamped on the floor. At last the wireless burst into sound. Judy Garland sang "Over the Rainbow".

"Turn it up! Turn it up!" James shouted.

Brian would hate that song for ever more. It was followed by dance-band music. The three walked aimlessly about the loft, picking up implements and off-cuts of leather, then casting them aside. And then, abruptly, the din below ended and the music blared on incongruously.

"Turn it off, for Christ's sake," James called. There was a weird silence, broken only by Peter Joseph's sobbing in the corner. Hughie emerged from the cowshed unsteadily and leaned back against the wall.

"Peter, put on the kettle and make a mug of tea for the sergeant," he called in a trembling, croaking voice. "And ask James if he has a cigarette to spare."

Brian sat upright in bed. He was sweating profusely. Bridie took his trembling hand and looked anxiously into his terror-stricken eyes.

"It's all right. It's all right!" she kept saying. "It was just a bad dream, a nightmare. Probably about the cattle."

He could not speak but shook his head fiercely. He could find no words, no explanation for the nameless fear that now sent a gnawing cold crawling through his body. Bridie wiped his brow and drew the blankets about him.

"I'll make you a nice hot drink to help you back to sleep," she said. He nodded but he could find no solace in her words. Something he could not name was causing him wrenching pain and great fear. The only thing he was certain of was that the pain had nothing to do with dead cattle.

Two days later Bridie came home from Cullyboe in a distressed state. She handed Brian a rolled-up newspaper. Her voice trembled.

"Holy Mother of God, such carnage! God have mercy on them all!"

Brian smoothed out the newspaper and read the headlines with growing incredulity and fear.

Germans Bomb Belfast
Over 500 Feared Dead
Many Thousands Homeless
Churches, Schools, Hospitals Razed
Fire-Brigades Sent from Éire

Brian looked anxiously to Bridie. The gnawing cold within him returned.

"Thank God it's the Easter holidays," she said. "At least Hannah will be out of there."

At the mention of his sister's name, Brian knew what was tormenting him.

"I'm going up to Sloe Hill," he whispered.

Bridie was momentarily stunned by his remark. By the time she realised what he was suggesting, he had already mounted his bike and was pedalling furiously down the road through a heavy drizzle.

He laboured up Caraher's Brae as the drizzle turned to driving rain which soaked his light clothing. Several times he nearly fell off the bicycle as he negotiated the rain-filled ruts. His heart pounded from both his exertions and his growing fear as he turned into the gates of Sloe Hill. He pounded on the front door and, on getting no answer, he raced down the haggard, calling his sister's name. The rain grew even heavier as he blundered in and out of outhouses calling for Hannah.

A towering figure in black oilskins stepped out from behind a haystack.

"She's not here," Winston Pollock said.

"Where – ?"

"In Belfast. She decided to stay on there for the holidays. Of all times –"

"Aunt Rose?" Brian panted. "Is Aunt – "

"She's gone to Belfast to look for her. I told her not to go but . . . women for you . . . There were no telephones working." He stabbed a pitchfork into a bundle of hay. Brian leaned against the haystack, burying his face in the damp hay.

"It's a bad business up there but I'm sure they'll be all right. Belfast is a big place." He raised the forkful of hay onto his shoulder and walked away in the direction of a field at whose gate hungry cattle stood lowing plaintively.

CHAPTER THIRTY

HANNAH JOHNSTON WAS ANGRY. SHE CLIMBED INTO BED QUICKLY AS SHE heard Hazel's footsteps on the stairs. She did not wish to talk to her cousin. Not after the way she'd been treated.

Hazel padded gently towards Hannah's door, not wishing to disturb her parents. She bent down to the keyhole.

"Coo-ee! Coo-ee! Coo-ee!

Cecil Burns is in love with ME!"

she whispered. Hannah said nothing. Hazel could not stifle a giggle and, in a moment of carelessness, she trod on the squeaky floorboard on the landing.

"Who's that?" her father called from the back room.

Serves her right, Hannah thought.

"It's only me, Daddy," Hazel answered. "I need a drink of water."

Drink of water. She needs a bucket of water poured all over her . . . Hannah gritted her teeth.

Why did I let her persuade me to stay here for the Easter holidays? "We'll go for a picnic on Cave Hill!" Hmph. We "accidentally" bumped into Cecil Burns outside the Café Rio. We decided to go to the pictures instead. Not to *The Wizard of Oz*, because Cecil had already seen it. With his mammy . . . No, we decided to go to a boring cowboy picture instead. Not that Miss Hazel Taylor saw much of it. She stayed in the back row wrapped around Cecil Burns while Hannah was dispatched six rows down and spent the entire duration trying to avoid the groping hand of a middle-aged man who timed his moves to coincide with the peaks of action on the screen . . .

When it came to the shootout at the bank, Hannah could take no more. She marched out to the foyer and waited there for Hazel and Cecil who were oblivious of her early exit . . .

Hazel had used her. There was no other word for it. She had needed an alibi. A picnic on Cave Hill with Hannah would be perfect. Sensible Hannah. "Maybe a sensible wee girl like Hannah will make you give up your wild ways," Hazel's father said. Used. They all used her.

She would teach them all a lesson, especially Hazel. It was only Easter Tuesday. She would go home to Sloe Hill tomorrow. She would find a way. And she might not come back at all. Never. Well, maybe. There was Mr Barden's Bible Class. Well . . . Mr Barden . . .

"Therefore saith the Lord God, behold I will stretch out mine hand upon the Philistines and I will cut off the Cherethims, and destroy the remnant of the sea coast . . . And I will execute great vengeance upon them with furious rebukes; and they shall know that I am the Lord, when I shall lay my vengeance upon them . . . when I shall lay my vengeance upon them . . . my vengeance . . . "

The steady drone overhead woke her. She lay rigid in the bed. It was a hostile, ominous noise. She knew she should move but she could not. The sirens began. Then the whistling sound . . . Hazel's father had just called out the girls' names when the first explosion went off. A series of explosions followed with great rapidity. The noise grew nearer all the time, growing to an unbearable pitch. Hannah clapped her hands over her ears and screamed. Then came the searing orange flash. The house shook and seemed to collapse in slow motion. She was falling, falling amid debris and furniture. Suddenly everything went black.

One side of Lambeth Gardens had been obliterated. Flames leaping from the ruined houses silhouetted the shattered walls against the glowing skyline. Further away the sky was illuminated by a series of flares drifting lazily to earth. Smoke drifted in billows through the shells of the houses. There was a total silence. It was as if she were watching a silent picture. For a terrifying moment, Hannah thought she was deaf until her ears popped open, as if a cork had been drawn simultaneously from each one. The noise was frightening – sirens blaring, screams of pain, names called out in desperation, lesser explosions from gas mains, howling from dogs. Amid all the din Hannah thought she heard Hazel's voice calling her name once but she could not locate the source of the call. She could not even distinguish the house she had been sleeping in.

It was still dark when she awoke. She lay underneath her own bed. The dust-laden air and the acrid smoke were choking her. She wormed her way out from under the bed. As her eyes grew accustomed to the unreal light, the scene about her defied comprehension.

She had to get out of this place. She would go home to Sloe Hill. There would be no bombs there. She would walk all the way if necessary. She staggered out of the debris which cut her bare feet and tore at her nightdress. She made her way with difficulty to Lambeth Avenue. The street was crowded with panic-stricken people who were equally intent on getting out of the city. Prams, bicycles and push-carts were pressed into service to carry the meagre belongings they had managed to rescue. Hannah grasped the collar of her nightdress tightly about her and stumbled onwards. A man staggered past her with a greyhound pup under each arm. A young woman with wildly staring eyes suddenly confronted Hannah.

"Milk," she cried weakly. "I need milk for my wean." Hannah looked down at the bundle in the woman's arms. A tiny baby wrapped in a tattered blanket, its face completely blown away. Hannah screamed and ran away. She stopped, clung to a lamppost and vomited violently. She felt weak and cold. Her nightdress, soaked in vomit, clung to her body. She had to keep going. A figure wheeling a bicycle paused to drape a rug around Hannah's shoulders before disappearing into the eerie light.

And then imperceptibly at first but gradually insinuating its way through the general clamour, the drone returned. The fleeing crowd paused momentarily to stare at each other in disbelief, before looking up to see the flares hanging in the orange sky almost directly overhead.

"The fuckin' bastards are back," a man screamed. "Get off the street! Get off the street!"

Hannah ran blindly along the pavement until she came to a little park. She veered into the park, past swings that swung crazily from the impact of an earlier exposion. She heard the whistle of the bombs and threw herself under a tree, covering her head with her arms.

There was an ear-splitting bang in the street she had just left. The last thing she heard was the tree above her split down the middle as if a giant hand had cleaved it in two with an axe . . .

Brian lay on his bed in the senior dormitory and idly followed the path of the plaster cracks in the ceiling overhead. Tommy Connolly paused at the foot of Brian's bed.

"Any news?"

Brian shook his head. Tommy shrugged his shoulders and moved on to the washroom. Brian had not wanted to return to school but Bridie had pointed out that there was little else he could do. Aunt Rose was still in Belfast, searching every day. The bodies of Hazel and her parents had been recovered and buried. There was still hope . . .

Father Declan breezed through the dormitory, tossing a letter on an occasional bed as he passed.

"*Brostaigh*, boys! Lights out in ten minutes!" He paused at Conor Maguire's bed to give him a cynical look before slinging a letter at him. "I never saw such an intimate letter from a granny, Maguire. Aren't ye lucky? And she has such a steady hand too!"

There were sniggers all around the dormitory as Father Declan disappeared into his room. He appeared again instantly with a bulky parcel under his arm. To Brian's surprise the parcel was deposited on his bed.

"Your aunt seems to know the bread-van driver well, Johnston. This came by special delivery this afternoon. I hope your aunt isn't like Maguire's granny . . . "

He went back into his room.

Brian tore the parcel open to the ribald comments of the other boys. Inside he found a large loaf of Bridie's soda bread and a pot of her damson jam. Wedged between the two was a single sheet of paper.

Dear Brian,

Just a note in haste to say there is still no news from Belfast. We are all praying away as I am sure you are too. God is good. Belfast is a terrible place at the minute. The people who were bombed are terrified to go back to the city. They say there are thousands out in the countryside. Some sleeping in ditches, God preserve them. Even in the city there are thousands homeless. Poor Aunt Rose came home exhausted. She says she will go back again in a few days. We live in hope.

I am taking a chance with Jemmy Conlon the breadman of getting this bit of food to you as I know how hungry you boys are. I hope you are well. It will not be long till the summer hoidays now. All are well in Drominarney, although it's hard on poor Hughie with all the cattle gone. It's very quiet up there now.

God bless you, Brian.

Your loving Aunt Bridie.

Brian folded the letter with a sigh.

"Has anyone got a knife?" he whispered after lights out. Tommy Connolly passed one up and ten ravenous boys feasted on soda bread and jam. Conor Maguire sank his teeth into a crudely cut heel of bread dripping with jam and with his mouth full he mumbled, "Good man, Breadman! I declare to God I love you nearly as much as my granny!"

The laughter rippled through the dormitory until a loud thump on the wall from within Father Declan's room interrupted them.

"Cut that out or you'll have extra study tomorrow night!" the voice said.

For the next two weeks Brian prayed as he had never prayed before. He was first into the chapel every morning and each time he had a few minutes to spare at meal-time or play hour, he slipped quietly into the chapel again. Please, God, please. Let her be found. Let her be all right. Please. Sacred Heart of Jesus, I place all my trust in thee . . . He prayed to St Anthony. Uncle James always prayed to St Anthony to help him find lost things . . . He thumbed through *The Garden of the Soul*, the prayerbook James had given him . . .

Quoniam in me speravit, liberabo eum. Protegam eum, quoniam cognovit nomen meum. Clamabit ad me, et ego exaudiam eum. Cum ipso sum in tribulatione. Eripiam eum et glorificabo eum . . .

Because he hath hoped in me, I will deliver him. I will protect him, because he hath known my name. He shall cry unto me and I will hear him. I am with him in trouble. I will deliver him and glorify him . . .

God is good . . .

Father Declan peered over his glasses.

"What's the matter, Maguire? No letters at all these days. Has granny found a new grandson?" As he turned away, Conor put his tongue out at the priest.

The letter floated towards Brian. He snatched at it and ripped the envelope open.

Dear Brian,

She's alive! She's in hospital – in a coma, but they hope she'll come out of it. No one knew who she was until a clergyman recognised her. She was found under a fallen tree in a park. Isn't it great news? We must keep praying. Glory be to God and his blessed Mother.

Your loving Aunt Bridie.

The tears ran down on the ink before he could fold the letter.

"What's wrong, Breadman?" Conor Maguire jibed. "Did your mot shoot you down too?"

"Shut your big mouth, will you," Tommy Connolly snapped.

"Lights out!" Father Declan called. "Mass at seven in the morning, boys. First day of the retreat."

Rose negotiated her way slowly along the hospital corridor. There were beds sited alternately on either side of the corridor and wherever a patient was receiving attention, it was difficult to squeeze past. The obvious suffering of the patients distressed her and she was relieved to move into the more spacious ward where Hannah lay.

Mr Barden was seated by Hannah's bed reading from his bible. He smiled at Rose and continued reading quietly . . .

"I waited patiently for the Lord and he inclined unto me and heard my cry. He brought me up also out of an horrible pit, out of the miry clay, and set my foot upon a rock and established my goings . . . "

Rose smiled at the pale, impassive face, willing it to make even the slightest reaction to Mr Barden's words. She gently combed Hannah's hair with her fingers. Ever since the day several weeks previously when Mr Barden had recognised Hannah as he toured the hospital giving solace to the bombing victims and their families, the girl had

229

maintained her serene appearance. The doctors could not give a prognosis. Her physical injuries were superficial. She was traumatised. That was the word they used but they could not forecast the ultimate outcome. Mr Barden had visited Hannah every day since . . .

"Yet the Lord thinketh up on me. Thou art my help and my deliverer. Make no tarrying, o my God."

He closed the book and looked intently at Hannah before speaking.

"I find great comfort in the psalms," he said. "Psalm 40 abounds with confidence in God and obedience to His will. That's what we all need at this time."

Rose nodded in agreement.

"It's just that sometimes it's difficult to maintain that confidence, Reverend," she said.

"We can only try," Mr Barden replied. "You know, I'm convinced she looks a wee bit brighter these last few days. I really am!"

"I'll be all right, Aunt Bridie. Sure, Aunt Rose is up and down to Belfast all the time." Brian watched anxiously for Rose's pony and trap as they stood outside Cullyboe church.

"Still – they've had an awful lot of bombing up there. Promise me you'll be careful."

"Of course I – here she is!"

Bridie sighed. Nothing would deter him from visiting his sister. She waved goodbye to Rose and Brian before turning back into the church to pray for their safe return.

"There she is now," Rose whispered, "sleeping away! You can talk to her, Brian. The doctors say we should talk to her all the time."

Brian found it hard to understand that his sister could have been "sleeping" like this for two months now.

"Now, Hannah, look who's come to see you today," Rose said brightly. She gestured to Brian to speak.

He spoke hesitantly. "Hello, Hannah. I'm – home from school. On holidays." There was no reaction from his sister. He looked to Rose.

"Just talk away there," she said. "I want to have a wee word with the sister."

Brian reached into his pocket and withdrew a small package

230

wrapped in crumpled newspaper. He took out the eggcup that Uncle James had restored after Lacy smashed it on the dormitory floor.

"D'ye mind the day I won this at the sports?" He slipped it into Hannah's hand and folded her fingers slowly around it. "I hope you still have yours. Mine had a bit of an accident but Uncle James fixed it."

He watched her face closely but there was still no reaction. He shrugged his shoulders. It was just an idea he had. He looked around the crowded ward. He realised only now, having walked through the streets of Belfast and having seen the patent suffering in the hospital ward and corridor, how horrific the bombings had been. At school they joked about the Jerries and wished the bombs had fallen on St Felim's. He smiled as he remembered Conor Maguire mimicking Father Colman.

"Point One. We have no school left. Point Two. Yis can all feck off home."

It was entirely different to come face to face with the reality of the bombing.

"Sister! Sister!" The young nurse's excited call woke him from his reverie. "Hannah's waking! Hannah's waking!"

Rose and the ward sister came running down the ward. Brian looked at his sister's face which a few moments before had been placid and expressionless.

Now her eyes were blinking. Her lips began to move very slowly. She spoke, almost inaudibly, but to the little group bent over her bed the words were quite clear.

"Where's Hazel? I heard her calling. Where's Hazel?"

CHAPTER THIRTY-ONE

THE CLOCK ABOVE THE BAR CHIMED ELEVEN O'CLOCK. THE THREE MEN standing at the bar were woken from a long silence by the reverberating chimes.

"God in heaven. Is that the time it is?"

"Not at all. Bridie has it set a quarter of an hour fast."

"Indeed I haven't!" Bridie laughed. "It's ten minutes slow. It needs winding. I must get Brian to – "

"Well, there's a great stretch in that day, surely. We could have got another hour at the hay."

"Ach. There's no need. The weather is set. There's no point in killing ourselves. Look at poor oul' Petesy Marry. What good did it do him, killin' himself?"

"What happened Petesy?" Bridie asked.

"He's away to the County Home, the poor oul' divil. Couldn't manage on his own any longer."

"He should have got a woman years ago – "

"Petesy? G'outa that! Petesy wouldn't let a woman next nor near the place."

"Wise bleddy man! That reminds me, Bridie. Have ye a bit of bacon at all? The missus told me to bring home a bit – "

"There's not a great selection, Charlie," Bridie said. "I'll see if I can get you a bit of collar." She moved down to the grocery end of the bar. "Didn't Petesy have a woman come in to look after the house?" she called back.

"That was a niece of his who came up from Dundalk in latter times. She insisted on coming. Petesy didn't want her at all. Said he could never find anything after she was gone!"

"Still, he had a tidy wee house for all his givin' out."

"What about his land?"

"Sure he only had ten acres worth talking about. He has it let out this three years – since the pains crippled him."

"Poor oul' divil. He'll only come out of the Home in a box."

"Somethin' will have to be done about that oul' mongrel of his. He won't leave the door of the house since Petesy went and he goes for anyone that comes near the place."

"There you are, Charlie." Bridie wrapped the bacon in a piece of newspaper.

"Begod, I got a bit of reading with it as well!"

He rooted in a fistful of coins for the money to pay her.

"There y'are! I'd better go before she sends the stormtroopers out lookin' for me!"

"Begod, Charlie, if it's that bad, you'd be better off movin' in with Petesy."

"Christ, no. Not that! I'll see ye, men."

"Good luck!"

Another long silence. Bridie rinsed Charlie's glass and thought again of what Brian had said to her earlier that afternoon before he had left for Drominarney . . .

"There's another spud in the pot, Brian, if you're – "

"No, thanks."

"Are you feeling all right? It's not like you to say no to another spud!"

"I'm grand. I'm full."

"I don't know." Bridie gave him an anxious look. "You don't seem to be yourself these last few days. Is there something bothering you?"

He looked away to the window. Hanratty passed by, his head hidden by the crate of bottles he carried on his shoulder. He sang "South of the Border" in a tuneless voice.

"Hasn't a note in his head, the craythur," Bridie laughed, "but it never stops him singing."

"I think I want to be a priest." Brian's voice dropped to a barely audible whisper. Bridie wiped her hands slowly and needlessly on her apron.

"You want to be a priest?"

He nodded.

"Are you sure?"

"I think so."

"How did this come about?"

"It kind of happened gradual, like. At first when the priest spoke to us, during the retreat. Then when Hannah was missing I prayed a lot for her – and I promised God I'd pay him back – if she was found . . . "

"You'd pay him back?"

He nodded again. "I promised I'd become a priest."

"Oh, Brian." Bridie rose from the table. "God doesn't think like that. He won't keep you to a promise made when you were worried about – "

"I know, but the more I think about it – I still want to. Are you cross with me?" he added quickly.

"Cross? Cross? Why would I be cross?" She moved around behind him and placed her hands gently on his head. "Isn't it the greatest blessing that can come on any family – to have a priest among them?"

"I thought you'd be disappointed that I won't be here – to help you with the shop and the bar."

"Don't you worry about that. Me and Hanratty will manage – as long as he doesn't sing too much!" They both laughed as they heard Hanratty begin to mangle another song in the bottling store. Brian got up to leave for Drominarney.

"I just want you to be sure that you're doing this for the right reason, Brian," Bridie said, reverting to a serious tone.

"Not paying back anyone . . . " He looked at her sheepishly.

"Anyway, you have another year at school to make your mind up." Brian nodded in agreement and moved to the door.

"And, Brian," she called after him, "your mother would be very proud of you . . . "

She watched him cycle up the hill towards Drominarney. As he cycled he tilted his head in a curious way. Bridie remembered how she used tease Sarah for doing the very same thing.

Man in the moon, Man in the moon,
Why did the dish run away with the spoon?

Rose Johnston observed, through a chink in the lace curtain of her bedroom window, the scene on the front lawn of Sloe Hill. Two figures sat under the chestnut tree, shaded from the warm September sun. Before them a linen tablecloth spread on the grass bore the remains of afternoon tea – part of Rose's best china tea service and a plate of fairy cakes and apple slices, practically untouched. The sunlight, filtering through the leaves of the chestnut tree, played on the blonde tresses of the girl and occasionally highlighted her pale features until she moved to avoid the sun's rays. On her lap lay a bundle of daisies which she had picked earlier. She worked intently at making a daisy-chain, splicing the narrow stems carefully with a thumbnail before threading another daisy through the splice. Rose smiled at the serious mien of her niece. Hannah could as easily be six years old as sixteen. In that very fact lay also the seeds of Rose's disquiet. Hannah was still a child, a very fragile child, and yet her relationship with the man who sat beside her puzzled Rose.

William Barden's sombre clerical garb contrasted with the pretty floral print dress of his companion. He leaned back against the broad tree trunk and read aloud from the scriptures, his expression alternating from animated to serious. Snatches of his reading came to Rose in the drowsy stillness of the afternoon.

"To every thing there is a season, and a time to every purpose . . . "

She had so much to be grateful to the Reverend Barden for. It was he who had recognised Hannah in the hospital, who had visited her daily while she was in a coma.

"A time to weep and a time to laugh . . . "

When Hannah had come home he came to see her as often as his circumstances permitted, which was at least once a week.

"A time to keep and a time to cast away . . . "

Hannah's rehabilitation had been slow and painful. The bouts of depression, the sleepless nights, the nightmares when sleep did come. There was no doubt that William Barden's visits lifted her spirits as nothing else could . . .

"He hath made everything beautiful in his time . . . "

And yet Rose sensed that the relationship was deeper than that of caring pastor and broken patient. But she was only sixteen – *a child* . . . he must be close to thirty, at least a dozen years older than Hannah.

"I know that whatsoever God doeth, it shall be forever . . . "

Rose had sounded out Winston's views on the relationship.

"Look, thon's a man of God and a loyal son of Ulster. There's no better than that. And he does the wean a power of good," Winston blurted. "What more can you ask, woman?"

Maybe he was right. Maybe.

Hannah reached over and dropped the completed daisy-chain over William Barden's head. She giggled as she arranged the chain properly. Mr Barden closed the book. They exchanged words and laughter.

Rose strained to hear what they said but the laughter defeated her attempt.

When Brian came home for his Christmas holidays in his final year at St Felim's, his mind was made up. He had talked with Father Colman about the priestly life. He was sure now. He had a vocation. He would serve God as a priest.

During the months that had elapsed since Brian had first told her of his intention to become a priest, Bridie had also dwelt a lot on the future. On Christmas evening, she and Brian visited Drominarney. Christmas Day was one of the few days in the year when the parlour was brought into use. It had been so for as long as Bridie could remember. Her father would procure an armful of logs on his return from Mass and would have a roaring fire going in the parlour by the time the family sat down to the Christmas dinner. She remembered how Sarah and herself had always beeen intrigued by the novelty of the parlour. It had a different smell to the other rooms – a mixture of mustiness and the wax polish which her mother had applied to the floor on Christmas Eve. Even the furniture was strange to the children. On the high mantelpiece – out of the children's reach – sat a collection of ornaments – a china shoe, a mug, a crucifix and a souvenir of Blackpool Tower.

It was that souvenir that Bridie fingered nervously now as she waited for Hughie to pour a whiskey for himself and James. Brian and the older children played in the kitchen while Josie put the younger ones to bed.

"I have something to say to you while we're alone," she began. "I'm selling up the place below – "

"The shop – the bar?" Hughie stood against the sideboard, his drink poised before his face.

"The whole lot."

"But why – where are you going?" Hughie lowered the glass without drinking from it.

"Well, Brian is going for the priesthood so he'll not be needing a bar and grocery – but he will need money." She paused for a few moments.

"And so will you, Hughie. To get back on your feet again."

"Ach, Bridie, you're talkin' nonsense," Hughie countered, his voice shaking.

"It isn't nonsense and you know it. You've had a hard blow and you have a young family to rear."

"I'll get by without – "

"You'll get by a lot easier with a bit of help."

"I can't take your money."

"You can take it as a loan, if it makes you feel better. Pay me back when you can."

"You haven't told us where you're going." James spoke for the first time as he lit a cigarette from the fire-ember he held in a tongs.

"I'm hoping to buy Petesy Marry's place."

"Petesy Marry's?"

"Aye. It's a nice tidy wee house – big enough for me and Brian. And there's a few acres with it. I think I'm more of a farmer than a publican at heart."

"That's a big step to take, Bridie." James poked the fire aimlessly with the tongs.

"I know that."

Hughie gulped his whiskey. James threw another log on the fire. It hissed and spat through a long silence.

"Look," Bridie said at last. "I'm close to fifty years of age. I'm not going to get married again – who would have me, for God's sake!"

"Gussie Fox would, for a start!"

She gave Hughie a withering look.

"All I want is a small place of my own," Bridie continued. "And I'm doing this for the family – for Brian, for you," she nodded towards Hughie, "although if you make another smart remark like your last one I might change my mind. And for myself."

James flicked the cigarette butt into the fire.

"Well, if it's what you really want to do – "

"It is – now open that bottle of sherry I brought and we'll have a drink on it."

She noticed the tears in Hughie's eyes, glistening in the light of the dancing flames.

"You're just like your father, Hughie McKevitt," Bridie said softly, laying her hand on his shoulder. "He always cried on Christmas Day."

CHAPTER THIRTY-TWO

BRIAN JOHNSTON WAS AWAKENED BY THE DAWN CHORUS FROM A SLEEP THAT had been brief and fitful. Pentecost Sunday 1949. In the half-light he could barely make out the time. Four-thirty. In another eight hours he would be Father Brian Johnston, a priest of God. Ordination Sunday, the culmination of seven years of preparation, of study and prayer. Seven years of memories . . .

The accents. That was what he remembered most about that first evening. Amid all the excited chatter, the variety of accents stood out – soft, melodic, harsh, guttural, drawling, rapid, musical, nasal. They represented every county in the country. Brian was totally bewildered. Even in St Felim's there had been no great range of accents – most of them were like his own and as such were accessible and comprehensible. But here . . . In the dining-hall that evening he found himself nodding and muttering to a chatty youth, whom he later discovered was from Kerry, but whose speech was totally unintelligible to Brian. It was his first indication that there was a world beyond Cullyboe, even beyond County Monaghan. Later he signed the book and the gate-keeper (whom everyone called Cerberus) assigned him a number – 42. He made his way with some difficulty to his room and had just begun unpacking his case when a sonorous voice rang out behind him.

"How's it goin', boy?" Slightly startled first by the strange voice and then by the huge-framed body from whence it came, Brian whispered a reply.

"All right."

"Arra cheer up! We'll be outa here in seven years – with good behaviour!" the giant chuckled. "Stevie O'Brien, Cloyne."

"Cloyne?"

"Cloyne. That's the diocese, like. You heard of Cloyne, surely? That's where Christy comes from."

"Christy?"

"Christy Ring. The hurler. Where have you come from at all, boy?"

"Monaghan – diocese of Clogher."

"Well that explains one thing anyway. I suppose you never had a *camán* or a *sliotar* in your hand?"

Brian shook his head in genuine ignorance.

"And you have a name, I suppose?"

"Sorry. Johnston. Brian Johnston."

Stevie extended a huge hand and, as he shook Brian's hand vigorously, a smile of delight spread all over his round face.

"Johnston! I don't believe it! Larry!" he bellowed towards the door while still holding Brian's hand.

A squat red-haired youth stood in the doorway sleeking his hair with a comb. Stevie made the introduction.

"Larry Rooney, diocese of Meath, God help us, meet Brian Johnston, diocese of Clogher, God help us even more!" Brian extricated his hand and offered it to Larry, who gave a friendly wink.

"At least Rooney here saw a hurling match," Stevie guffawed. A pause. Then, "Well, do ye not get it?" Stevie asked.

"Get what?" Larry was as puzzled as Brian.

Stevie tossed his great head upwards with a sigh.

"Johnston, Rooney and O'Brien! 'Tis close enough, like! Johnston, Rooney and O'Brien! The college bakers!" He exploded into laughter at his own humour. Brian and Larry joined in, caught by the infectious good humour of the jovial Corkman. From that first day for the duration of their stay in the college, they were "immediates." They sat beside each other, prayed beside each other, walked with each other. They were constant companions who over the succeeding years grew as close as brothers. For Brian, Stevie and Larry were the brothers he never had and he called them Big Brother and Little Brother respectively.

The unpacking was almost complete. Stevie's eye fell on the eggcup sitting on the shelf. He picked up the eggcup.

"Come on now, boy! What's this? Mammy isn't going to be bringing you the boiled egg and toast in bed from now on!"

Brian snatched it back from him.

"It's just – a souvenir – to remind me of – things."

"'Tis all right, boy. I know how you feel. I have two lovely duck feathers inside in my own room – just to remind me of home. I plucked them off the duck on the way out the gate this morning." For the first time his voice dropped.

"How about you, Larry?" Brian asked.

"Oh aye!" He suddenly whipped a five-pound note from his breast pocket. "I plucked this from the old man when he was saying goodbye!"

The room was bathed in the intense early light of a June morning. Everything except the clothes he would wear and his toiletries was packed neatly into two battered suitcases. A bundle of books, neatly stacked and secured by a stout leather belt, sat on the chair beside him. Brian smiled as he remembered how often the dean had berated him over the years for the untidiness of his room. He should see it now! The bundle of books told the story of his academic progress in the college – poetry anthologies, Latin and Greek texts, church history, theological tracts, a collection of scripture commentaries . . .

His first sermon before the Professor of Sacred Eloquence had been a disaster. He had been even more nervous than on his first appearance on stage in the Aula Maxima as a non-speaking servant in *The Merchant of Venice*. His sermon was for the Feast of the Annunciation. He prepared assiduously and spoke of promise, of reconciliation, of love. The Professor was scathing in his criticism.

"Basic ideas, sound, Mr Johnston, but the delivery was wooden, monotonous. No life, no colour, no variety. No engagement of your congregation. You'll have them coughing and dreaming within minutes of your opening. You're a human being, not a machine, Mr Johnston. Talk like one!"

Stevie staggered through a sermon on the Epiphany. The Professor's voice thundered from the back of the oratory.

"Your sermon had two merits, Mr O'Brien. It was audible and it was brief."

"Jakus, what did he want?" Stevie said after dinner. "I never spoke for more than five minutes on any topic – except last year's All-Ireland Final!"

Larry Rooney raised titters and giggles throughout the oratory with his sermon on the Parable of the Talents. His droll humour was lost on the Professor.

"Pure theatre, Mr Rooney, but for another stage, perhaps?"

"Yerra, what would ye know about talents up in Meath?" Stevie taunted him that evening.

"Know this, southern heathen!" Larry wagged a finger at him in mock threat. "Our appointed time is at hand, when the Lord shall see that it is right and fitting that the men from the Royal County of Meath shall inherit the earth – or at least Croke Park – and shall bear on high the spoils of a September day's battle – the Sam Maguire Cup!"

"Hah!" Stevie threw his great arms out in supplication. "That'll be the day! And here comes Cardinal Stevie O'Brien to throw in the ball! Ye have as much chance!"

Stevie lived for hurling and football. He was first out on the pitch at every recreation period in all weathers. He was not a polished athlete but his fearlessness and wholehearted attitude made him a priceless asset to diocesan and class teams. Larry Rooney was a useful corner-forward and, while Brian was usually selected on the class football team, he often felt it was just to make up the numbers. He enjoyed the games but lacked Stevie's passion and Larry's guile. In summer each diocese had its own croquet pitch and in his third year Brian had the distinction of leading Clogher to victory in the inter-diocesan croquet championship. He savoured the victory if only to tease Stevie, who was not impressed.

"Arrah, a game for ould mollies – and reverend mothers!" he announced, having failed to master its subtlety and delicacy of touch. In frustration, he swung the mallet in hurling style and skewed the ball down the gravelled path where it cracked against the dean's ankle . . .

Stevie carried his full-blooded approach on the pitch into his priestly life. During a Rubrics lesson his imparting of a blessing was sharply criticised.

"Please, Mr O'Brien. You are not fetching a ball and turning to set off on a solo run!" The Professor aped Stevie's flailing arm movements and ungainly walk to the amusement of the other students. "Where is the solemnity? The gravity of movement?"

Brian felt for Stevie in his embarrassment. He had experienced a similar moment himself one morning during a theology lecture when he was "called" to expound on a particular point. He had failed to read up on it the previous night and he limped through the "call" on a meagre supply of prompts from his neighbours.

"More attention to your studies and less to the sportsfield, perhaps, Mr Johnston?" was the Professor's acid response.

It was not the sportsfield that had distracted Brian but a letter he had received the previous day from Bridie.

Drumacool
Cullyboe
March 3rd 1946

Dear Brian,

Just a short note, hoping this finds you well. Unfortunately things are not good here. James is in the County Hospital with bad pneumonia. The doctor says his chest is in an awful condition with all the cigarettes. His breathing is very bad. Please pray for him and ask your friends to do likewise. There is great power in prayer, especially in the prayer of holy young men like yourselves.

The rest of us are all right. Hughie has three big fields in flax this year. He says he will need your help with it next August. There is great demand for the flax after the war and there is a good price for it. I am glad for Hughie. He had his bad times.

Please pray for James.

Your loving aunt,

Bridie.

Brian could not concentrate on any study that evening. James must not die. He could not imagine Drominarney without James. Every time Brian tried to settle down to study, images of the gentle James drifted across the pages of his theology text – crouched over a saddle that needed stitching, cigarette cocked at an angle between his lips; playing with his wireless like a child with a Christmas toy; treating Brian and Bridie to high tea in Castleblayney after a day's shopping;

sitting on a headland on a Sunday after Mass leafing through his prayer-book . . . In the event, James pulled through but he was a much weaker man – "failed a terror," as a neighbour put it – and his working life was effectively over.

Brian looked across at the desk. How many hours had he spent there? For all the diversion that games and the subsequent banter offered, lectures and study were the students' main concern and there were long hours of silent study each evening when theology, philosophy or church history texts were their sole companions. For a time Stevie smuggled in occasional copies of The *Cork Examiner*. His main interest was the results of local games, while Brian and Larry devoured the rest of the paper for news of the war. The smuggling stopped, however, when copies of the paper were discovered and Stevie spent an anxious hour in the dean's office. Before he went in to face the dean, Stevie was terrified. It was the first time Brian had seen him like that – nervous and fearful for his future. Attempts to humour him were fruitless.

"Don't worry, Stevie," Larry jibed, "for punishment, you'll probably have to read the news section!"

To Stevie's relief he was let off with a severe caution. Later he confided the source of his fear to Brian.

"'Twould kill Mother to have a second spoiled priest in the family."

It transpired that Stevie's older brother had left the college after two years and was now languishing at home, a "spoiled priest." He had left simply because he knew he did not have a vocation – a brave decision, but not recognised as such at home.

A vocation. Do you truly have a vocation? Are you prepared to pay the price? The questions were put to them during a retreat at the start of each term. Brian had never wavered. At times the length of the course and the demanding nature of the material to be studied tested his perseverance but he knew he would win out in the end. So, too, would Stevie, he reckoned. Stevie would make an excellent pastor. For all his ebullience, he was a caring man, generous in spirit and possessing a deep understanding of and interest in people. And Larry? Larry always claimed that the priesthood was only his second choice. "I want to own and train racehorses but you need money for that game! Maybe I'll get a rich parish that will pay good offerings!"

244

Five past six. Another twenty-five minutes before the rising bell – that bell that had regulated their lives over the previous seven years. He picked up *The Garden of the Soul* and thumbed through it for morning prayers, saying a silent prayer for the book's donor, Uncle James, as he did so. A tiny newspaper cutting fluttered from the leaves of the book onto the counterpane. He picked it up and read it again.

"The wedding took place recently of the Reverend William Barden of Ballymena, County Antrim and Miss Hannah Johnston of Sloe Hill, County Armagh. Mr Barden has recently been appointed Rector of Castlestewart Parish in County Tyrone and will take up duties there next month."

Aunt Bridie had sent the cutting and had written the wedding date underneath – 14th March 1947. She added in the accompanying letter that "Rose was not at all approving, considering that Hannah was not yet twenty-one. I must say I would have to agree with her. She is little more than a child but no one was going to stop her . . . "

Brian had been startled by the news. Since he had opted for the priesthood he had seen little of his sister – maybe once each summer. She had become very withdrawn since the bombing incident and had generally shunned company – Mr Barden being the only exception. Her reaction to Brian's decision to enter the priesthood had been decidedly cool – there were references to "joining the Pope's army" and attending "the training camp." Now unexpectedly, his sister had married. He had not even been invited to the wedding. That had upset him greatly. If an invitation had come he would not have been able to attend but that was not the point. It would have been an acknowledgement of his existence in Hannah's life. Their paths, once parallel, now seemed to diverge at an ever-increasing angle. He had written a brief congratulatory letter to Hannah and William Barden. Two years later he wrote another letter – to congratulate them on the birth of their daughter.

The rising bell jolted him from his reverie. Even after seven years that bell could still startle him. He leaped out of bed instantly. A busy day lay ahead.

CHAPTER THIRTY-THREE

THE SWELL OF THE GREAT ORGAN FILLED THE COLLEGE CHAPEL AS THE procession moved slowly up the aisle. Fifty-three young men, bearing themselves with immense pride and solemnity. Brian was towards the rear of the procession. He looked straight ahead towards the altar, as they had been instructed to do, but halfway up the aisle he caught a glimpse of Bridie, her face aglow under a wide-brimmed navy hat. He was afforded a wider view than the others in the procession through the absence of a partner. He walked alone.

Eight months previously when the rising bell had echoed through a misty November morning, Brian struggled through the beginning of an unwelcoming day. There was a damp chill in the air. He checked to see if he had left his window open but it was shut tight. He shivered. Something more than the cold was disturbing him. He paused. The muffled sounds of early morning came to him – a slap of water, a scrape of chair on the floor, a snatch of a tune hummed in front of a mirror. He knew now what disturbed him. There was no sound from Stevie's room, normally the noisiest room on the corridor. Brian thumped the wall three times with his fist. No reply. Come on, Stevie. He half expected to hear Big Brother's snoring in response. Nothing. He moved quickly. When he threw Stevie's door open a new chill ran down his spine. Stevie lay motionless, his eyes rolled upwards, his breathing rapid and distressed.

"Stevie!" Brian's anguished cry reverberated through the house. "Stevie! What's wrong?"

Two days later Stevie was dead. Meningitis. Men-in-gi-tis. Brian hated every syllable of that ugly word. He lay in bed and gripped the bar of the bedstead fiercely. In the stillness he could hear Larry sobbing in the next room. Why? Why take Stevie now, on the threshold of the priesthood? He would have made a marvellous priest. Why take the worker before he had even entered the vineyard? He could not rationalise it. He could not accept it. Why? Why? Why? Why should he continue in the priesthood himself?

Stevie was buried in the college grounds on a miserably grey November day. Six of his fellow students, Brian and Larry at the front, bore the coffin down the yew-lined walk, while the entire student body sang the *Benedictus*. Brian gripped Larry's shoulder tightly as he felt the shudders running through his body. At the graveside a little knot of family mourners stood across the grave from Brian. A tiny bird-like woman dressed entirely in black fingered her beads as she stared in disbelief at the grave now being filled in. At the end of the ceremony Brian introduced himself and Larry.

"Stevie was always talking about you, Mrs O'Brien," he began awkwardly. Larry nodded in agreement. "And I want to tell you that he would have made a great priest. He had all the makings of one. He was just so genuinely good. And he gave us so many laughs . . . "

The little figure nodded her head gently as her lips continued to pray the rosary. A tall, slim young woman offered her hand to Brian.

"I'm Sheila, Stevie's sister. He spoke a lot about yourself and Larry – Johnston, Rooney and O'Brien."

Brian smiled. "The college bakers," he added.

As they moved back to the college, Sheila walked with Brian.

"It was very good of you to talk like that about Stevie to Mother."

"It was only the truth. I can't understand why he was taken. He had so much to give."

Sheila was in many ways unlike her brother. She was elegant and graceful in her movement, quiet and soft-spoken, but she had the same open face as Stevie and an identical winning smile. Brian felt at ease in her company although he had practically no experience of the company of women of his own age. She was a primary teacher in Dublin and she amused Brian over lunch in the college with tales of the eight-year-olds in her charge. Meeting her softened the blow of Stevie's death for Brian and he dared to hope their friendship would continue.

"Maybe you'll write sometime – just to keep in touch," he said awkwardly when it was time to leave.

"Of course!" Her smile might well have been Stevie's, saying, "Why wouldn't I, boy?"

Tu es sacerdos in aeternum. The words echoed through his head as the procession began its return down the aisle. A priest forever. He felt a great happiness within him. He could equally feel great waves of warmth and pride radiating from the congregation as the procession passed among them. Outside in the sun-washed square the crowd broke into excited family groups. Brian was thrilled to have not only Bridie and James present, but also Hughie and Josie whom he had not expected to come.

"Why wouldn't we come?" Hughie exclaimed as he struggled with a tight-fitting shirt collar. "It's not every day we have a priest in the family! James booked Tommy Murray's hackney for the day so we're going up to Dublin for a big feed as soon as you're ready!"

Brian gave his first blessing to his family. As Bridie knelt before him on the grassy incline, her proud, lined face reminded him of the many sacrifices she had made for him, the pain she had suffered, the loneliness she must have endured. James was next. Even though it was a pleasant June day he wore a heavy, belted overcoat. Brian noticed how frail he had become. James rose unsteadily from the kneeling position and reached into his pocket for a packet of cigarettes. He lit one and drew deeply on it. As he gave the blessings, Brian could not explain the recurring blurred image of a man falling down between drills of potatoes which continued to distract him . . .

Hughie tugged at his shirt collar. "Can I take this off now? The last time I wore one of these curse-of-God collars was at Peter Joseph's confirmation and I near died of suffocation."

"No, wait a minute!" Josie rooted in her handbag. "We have to take the photograph. Get into a line there, let ye – "

"Here, let me take the photo so you can all be in it!"

Josie was startled by the sudden intrusion of the tall woman in the tailored black suit. There was a momentary awkward silence before Brian leaped forward.

"I'm sorry! This is Sheila O'Brien, Stevie's sister." He turned to

Bridie. "You remember Stevie, that died." Bridie nodded. Brian introduced Sheila to his family, before she took the photograph.

"Well, I don't know about youse, but I'm starvin'," Hughie announced. "Where's Tommy Murray gone anyway?"

"Would you like to join us for a meal, Sheila?" James asked.

"No thanks. I'm with a friend. I just came down to wish Brian and Larry well –" she nodded in the direction of the Rooney family group " – just for Stevie's sake."

"There's Tommy," Hughie called. "Come on, will youse?"

"Nothing will come between Hughie McKevitt and his food," Josie laughed. They moved to follow Hughie while Brian lingered with Sheila.

"You had better go. Your Aunt Bridie is giving me suspicious looks," Sheila said softly.

"It was very good of you to come. I didn't think – "

"Stevie would want me to come. Now, Father Johnston, may I have your blessing?"

"Of-of course." The request and the unaccustomed title made him nervous.

"How do you feel?" Sheila asked.

"Happy, proud, nervous."

"Nervous?"

"Well it's been like a rehearsal up to now! It's the real thing from now on."

"Where are you going?"

"To a parish called Drumcash. There's an old PP there, Father Maguire. That's all I know about the place."

He caught sight of Bridie glancing back at him.

"What about you?" he asked.

"I'll be at home for the summer. Mother is not well."

"I hope we can keep in touch." He felt guilty the moment he heard himself say the words, but her smile reassured him.

"Now, you had better go or Hughie will go without you!" She moved away quickly.

"Thanks," he called. "Thanks again!" He watched as she slipped away through the crowds.

"Father Johnston! Put it there, old son!" Larry Rooney stood before him, grinning broadly as he extended his hand in congratulation.

"Father Rooney! Did we ever think we'd see the day?" Brian laughed.

"Well of course I never had any doubts, personally." Larry affected a serious tone. "Speaking of doubts," he whispered, "did you meet herself?"

Brian was taken aback by his remark.

"Oh now, it's just as well they're shipping you off to the back end of Monaghan!" Larry sensed his friend's unease. "I'm only joking, Brian, for God's sake. What's the name of that place again – Drumcash?"

Brian nodded.

"Well named – I hope!" Larry chuckled. "Good for you. I'm just after learning that I'm stuck with a crusty old tightwad of a parish priest. I'll be waiting for a racehorse, I think!"

Brian gave him a playful punch.

"Will you give me a hand with my case and stuff? I have an uncle who is dying of starvation."

Brian had a week's leave at home before taking up his duties in Drumcash. On the morning after his ordination he presented himself at the presbytery in Cullyboe, half an hour before he would say his first Mass. The housekeeper showed him to the dining-room where Canon McKenna was having breakfast. He was bent over the table, his slow laboured movements making the topping of his boiled egg a difficult task.

Brian stood self-consciously by the table. There was no invitation to be seated.

"You're cutting it a bit fine for nine-o'clock Mass," the Canon said at last.

Brian checked his watch in surprise.

"Celebrating the sacrifice of the Mass is a great privilege, a gift given to the few. It requires preparation. I have spent a half-hour in prayer and meditation before Mass every morning of my priestly life." His trembling hand caused the egg yolk to drip on the tablecloth.

"That's a great record indeed," Brian said.

"'Tis no record at all. The priesthood is about privilege. It demands commitment. Don't ever forget that!" He chomped on a piece of toast. The egg yolk dribbled from the corner of his mouth.

"So Sarah McKevitt's boy turned out a priest. God's ways are strange. A pity she is not here to see this day." He slurped a drink of tea. "I did my best to talk her out of that marriage but she was too headstrong. I hope you're not like that," he added quickly. Brian was growing more uncomfortable as he stood watching the old man's feeble movements. He attempted to change the subject.

"Would you like me to take the half-seven Mass this week?"

"Indeed not. The half-seven Mass is my Mass." He scraped some butter along a slice of toast. "If she had listened to me, she would be here this day."

"We will never know that," Brian countered weakly. He felt the beads of perspiration gather on his forehead.

"Of course we will." The Canon raised his head and stared up at Brian. As he did so, he shaved the dribble of egg yolk from his chin and spread it on his toast. "Wasn't it guilt over her decision that caused her to – that caused her death in the end? Would you pour out a cup of tea. It'll save me sending for Mrs Hughes."

Brian had to steady the pouring hand with his left hand. This bitter old man had thrown his mind into turmoil. He excused himself and made his way to the church to vest for Mass. What should have been an occasion of great joy had been clouded by confusion and doubt. He walked onto the altar with the Canon's words echoing in his head – "Wasn't it the guilt that caused her death? . . . "

The celebration of the Mass calmed him and when he went to meet family and parishioners afterwards he could genuinely participate in their rejoicing. Master Lennon formed the fifth and sixth class students from Drominarney school into a guard of honour in the churchyard. He took Brian's hand warmly in both of his and his eyes danced with pride.

"D'ye mind the time we took you to the Eucharistic Congress, Brian? The seeds of a vocation were surely sown that day, thanks be to God!"

Bridie eventually rescued him from the well-wishers.

"You'll meet them all again tonight," she said. "Josie's having a celebration in Drominarney. After I get your breakfast I have to go over and give her a hand. You'll manage on your own for a while?"

"Of course," he smiled. He needed to talk with Bridie but this was not the time.

"It's a brave while since I heard so much music and laughter in this house," James said as he watched the dancers swirl around the kitchen to the music of Josie McKevitt, her brother Matt and her son Peter Joseph.

"It's good to hear and to see," Brian agreed. It had been a totally enjoyable evening for him. It seemed as if the entire community had crammed into the kitchen. Some of them he barely knew but he could sense the pride and happiness in each one of them. James lit a cigarette.

"So you're off to Drumcash?"

"Aye, do you know much about it?"

"Not much. They used to have a middlin' football team. Pete had a saying about them.

'A bunch of ould mollies from the bog of Drumcash

Their bellies are full of turnips and mash'."

They stepped back smartly out of the path of a wildly swinging dancing couple.

Pete.

"Didn't Pete live in a wee white house?" Brian asked.

"Aye. When he had the weakness. It took him in the end. He was mad about the football."

The blurred picture came into clear focus.

"Aunt Bridie! Aunt Bridie! Uncle Pete fell down in the spuds . . . "

"Come on out of here before we're trampled on!" James said. "I've something to show you."

Brian followed his uncle across the haggard and up the stairs to the loft. A hay-scented breeze caused him to linger at the top of the stairs and savour the fragile warmth of a June evening. He noticed how James fought for his breath as he struggled to open the door.

"There y'are. You can amaze them all in Drumcash now!"

A brand-new electric wireless sat on the workbench.

"I doubt if, no more than ourselves, they have the electricity over there yet, but it's on the way and you'll be the first to have an electric wireless."

Brian was overwhelmed by the gift.

"Uncle James! It's beautiful – but it's too much. I can't – "

"You can and you will. I mind the times you spent listening to them concerts on the battery radio when I wanted to hear Bing Crosby. Well you can listen away now – as long as the PP doesn't object!"

Brian took his uncle's hand and shook it warmly. Neither of them knew of any stronger way to express their feelings.

"Thanks," Brian whispered.

"James is one of the best," Bridie said as they cycled home at midnight. "Your mother was mad about him – and he about her. I don't know what we'd do if anything happened – "

"Aunt Bridie, what happened to my mother?"

The whirr of the bicycle wheels accentuated Bridie's silence.

"You told me she was taken with an illness, but Canon McKenna seemed to suggest – "

"That man! Can't he let her rest in peace?"

She braked suddenly, leaped from the bicycle and then began to wheel it slowly along Flaxpool Lane. Brian did likewise.

"What happened – "

"What I told you was true and it was all you needed to know as a child. She had an illness – her mind wasn't right and she couldn't manage. She – drowned in the river. No one will ever know how but the Canon decided that she took her own life and refused her burial in the graveyard."

"So where – "

"In the children's graveyard. I can show you tomorrow. There's just a lump of rock . . . " They both stopped.

"But I was led to believe she was buried with my father."

"I know. I'm sorry, Brian. I only said that to protect you from hurt. There was enough hurt in our family at the time . . . "

They began walking again. Wisps of mist floated through the fields on either side of them. Bridie remembered a summer evening when Jimmy McCabe gave her a lift home on the padded crossbar of his bicycle. "The McKevitts always liked to travel in style . . . "

She looked across at Brian.

"Don't let it spoil your day, Brian. Your mother is in heaven, no matter where her earthly remains are – "

"I want you to show me the grave now."

"Now?"

"Yes, now. Please, Aunt Bridie?"

They resumed cycling. No words were spoken until they reached the little field opposite Cullyboe cemetery. The mist-covered grass clung to their legs as they waded through the field. In the moonlight

Brian could discern rough stones jutting crazily through the overgrown grass.

"No one ever touches this place," Bridie said as she made her way to the furthest corner of the field. She paused, uncertain for a moment, then pointed to a particular stone.

"That's it. That's where she's buried."

Brian dropped to his knees, careless of the sodden grass that soaked his trousers. He knelt there in silent prayer for some minutes. Tomorrow he would borrow a scythe from Hughie and put some order on this neglected place.

CHAPTER THIRTY-FOUR

THE HACKNEY CAR ROLLED SLOWLY THROUGH THE VILLAGE OF DRUMCASH. Tommy Murray interrupted the aimless tune he had been whistling through a gap in his teeth.

"Begor, Father Brian, you'll not be run off your feet in this place! Are they all in the cemetery?"

Brian laughed. Tommy had a point. It was mid-afternoon on a Sunday in June. There was not a human being to be seen on the street. An old sheepdog lay curled up in a shop doorway. Further down the street another sheepdog ambled across the road before settling in the shade of a house gable.

"Maybe they're all praying. Holy Hour," Brian suggested.

"Kicking football, more likely," Tommy grunted. "They're a terror for the football round here. Bloody savages, too – if you'll pardon my language!"

"Their bellies are full of turnips and mash!"

"Oh, you heard? Full of porter too, before every match, if you ask me!"

The church dominated the centre of the village. The silver railings, newly painted and gleaming in the June sun, seemed to stretch for half the street's length. Tommy swung the car through the gateway towards the two-storey grey house that stood to the right of the church. When they got out of the car, Tommy's suspicions were confirmed. A continuous raucous din came from the far end of the village.

255

"Some poor divils experiencing Drumcash hospitality," he said with a toss of his head, as he untied Brian's bicycle from the boot of the car.

Three rings of the doorbell brought no reply. "Himself must be at the match too," Tommy surmised. Brian was about to explore the rear entrance when the front door was suddenly thrown open.

"My apologies. Lizzie has the afternoon off and I was taking a post-prandial rest. You'll be Father Johnston?" Father Maguire offered his hand in welcome. The parish priest's corpulent figure almost filled the doorway. His great round head, bald on top with a collar of white hair from ear to ear, reminded Brian instantly of the advertisement for "Mac's Smile" blades. His rosy cheeks were traversed by a network of broken veins. "Come in, come in. You can park your luggage in the hallway for now." Tommy gave Brian a hand with the assorted baggage.

"That's a very fancy-looking wireless," the parish priest mused.

"It's an electric one."

"Well, you'll have to wait another year at least to use it. We're on the list for next summer."

Tommy took his leave of Brian.

"You'd like some tea after your journey?"

"I can help myself if you show me – "

"Not at all. The kettle's on the Aga and I asked Lizzie to leave the table set."

The dining-room table was neatly set for two with covered plates of fairy cakes and sandwiches sitting between the two settings.

"God bless Lizzie. She's a treasure. Make yourself comfortable there while I boil up the kettle."

Brian sank into one of the two leather armchairs that dwarfed the fireplace. The room was over-furnished. A huge mahogany sideboard, laden with an extensive silver service, ran almost the length of one wall. A Queen Anne china cabinet and a chiffonier stood on either side of the door against the opposite wall. A grandfather clock straddled one corner, its ticking seeming to grow more laboured with each swing of the pendulum.

"Now, sit over! sit over!" Father Maguire carried a cosy-covered pot of tea into the room.

"There's only two things that concern the people of this parish –

256

saving their souls and football. A good proportion wouldn't put them in that order either." Father Maguire offered the sandwiches to Brian. "Lizzie'll be annoyed if we don't finish these off." Brian took two sandwiches. "That's the spirit. I can see yourself and Lizzie getting on well," the older man mumbled through a mouthful of fairy cake.

"They're decent hard-working people in Drumcash – but that's not to say they are perfect. Far from it. They're proud, arrogant, secretive, calumnious, petty, vindictive, covetous, conniving, hypocritical. Aye and they're fearful, guilt-ridden, anxious, worried and doubting too. Sometimes they're even happy – when they win at the football." He gave an impish grin. "In other words, they're human – full of all the frailty and fragility that flesh and blood and bone try to contain. That's the best piece of advice I can give to a young man like yourself starting out in the priesthood. When I was your age I thought I had it all worked out, but I soon found out that textbooks and scholarly lectures were a pale imitation of life. If you remember that, you'll save yourself a lot of disillusionment. And in case I didn't say it before now – welcome to Drumcash!"

That welcome was gradually extended by the parishioners of Drumcash in the ensuing weeks. Brian found them shy and reserved in his presence, but there was reserve on his side too. He did not wish to appear too forward with them and preferred to let time take its course. Like the River Gash that meandered crazily through the adjoining townlands, he would find his own path, take his own level. For all that, it surprised him that he was kept so busy. Daily Mass, while at the core of his priestly life, was scheduled, predictable. It was the unscheduled events that ate into his day – a call to the sick, visitation of homes, a funeral, dealing with family quarrels, tragedies. He would never forget the first tragedy he had encountered. Matty McDermott was gored to death by a bull while working for a local farmer. Brian felt utterly helpless as he watched Dr McRory battle to save Matty's life as he lay at the headland on a bed of bracken. He administered the last rites and mouthed prayer as the life ebbed from Matty's crushed and bleeding body – that same body which, strong and wiry, had withstood the buffeting of championship football on the previous Sunday. It fell to Brian to break the news of Matty's death to his young wife, Mary. Helplessness welled up in him as she stood in the kitchen of their cottage, her baby, the youngest of four children, in her arms, her face contorted in a mixture of incredulity and deep anguish. Brian longed to hold her in his arms as she began to sob gently but that was not given to him to do. A neighbouring woman would console Mary. The river would find its own path.

Nothing in his training could have prepared Brian for confessions, either in terms of the variety of guilt and anxiety that troubled people or in the long hours of confinement in the confessional. On his first Saturday in Drumcash, he turned up punctually at noon for morning confessions. There would be another session at six o'clock that evening. He seated himself in the confessional and waited nervously for the first penitent. It was Katie McShane. She was in an equally nervous and agitated state. Brian did his best to put her at ease. Just as she was about to enunciate her sins, the Angelus bell began booming directly overhead, drowning Katie's outpouring completely. Brian had not the nerve to ask her to repeat her sins and dismissed her with a penance of three Hail Marys. Katie obviously spread the word that afternoon that the new curate was a "cushy" man in confession. When the evening session came, there were five pews full of penitents waiting for him. This time he waited for the Angelus bell to ring before entering the confessional, but the damage had been done. His reputation as a confessor had been well established.

Early in September he received a letter in the post. He did not instantly recognise the handwriting but was delighted to discover that it was from Larry Rooney.

Dear Brian,

How is it going? Has the new curate captivated the hearts of all in Drumcash? I hope so. As for myself, the less said the better. Life is full of trials and the biggest of them is one Father Collins PP . . .

Anyway the big news is that the men from the Royal County have finally made it to the All-Ireland Final. You did notice? Stevie O'Brien should be living at this hour! I expect we'll make short work of the Cavan fellows. Will you join me for this historic occasion? I'll be outside Barry's Hotel at two o'clock. I hope you can make it!

Regards
Larry.

When Brian disembarked from the train at Amiens Street, he attached himself to a group of Cavan supporters who assured him they would guide him to Barry's Hotel. They were in high spirits as they sauntered up Talbot Street.

"We'll look after you, Father. Just give the boys in blue your blessing – "

"They don't need any blessing. Just stay around to give the last rites to them Meath fellas!"

"Yahoo!"

"Come on Cavan!"

A car festooned with green and gold banners passed them by.

"Go home and ate your dinner outa the drawer!" a voice bellowed from within.

The Cavan supporters swarmed around the car but it sped away from them to a chorus of ribald comments.

They regrouped and trudged sheepishly back to Brian.

"Beggin' your pardon, Father, but you can see the kind of savages we're up against today!"

The incident set the tone for the day – exuberance, banter, good humour, occasional taunting. The city streets resounded with the voices of people who were enjoying a very special day out. Brian found it easy to enter into the spirit of the day. He recalled his only previous visit to Dublin as an awe-struck seven-year-old on an even more celebratory occasion. It brought home to him how sheltered his life had been – Drominarney, boarding school, seminary, Drumcash – his world had been a very prescribed one.

"There y'are, Father! Will you be all right now?" An excited crowd was already milling around Barry's Hotel. Brian assured his Cavan minders that he would manage and wished them well. They left him with a chorus of wild whoops.

"Up Cornafean! Come on the Gunner Brady!"

There was no sign of Larry but Brian was content to mingle with the crowd and savour the atmosphere.

"Hello, Brian!" The soft musical voice was unmistakable. He turned to greet Sheila O'Brien.

"Sheila! Where – "

"Larry obviously wrote to you too." She laughed nervously as she offered him her hand in greeting. Brian suddenly felt embarrassed and self-conscious in the middle of the swelling throng. He was delighted to see Sheila again but not here, not where he felt the eyes of hundreds were focused on them. They exchanged trivial remarks awkwardly before Larry came to their rescue.

"Well I see you two found each other!" he chirped, putting an arm about each of them.

"You are incorrigible, Larry Rooney," Sheila scolded him.

"Not at all! Isn't this a historic day and shouldn't Johnston, Rooney and O'Brien be re-united for it? Or as near as dammit?" he added with a hint of sadness. He turned to Brian. "Well, what do you think now?" He leaned forward and whispered, "Any doubts about your vocation, Mr Johnston?"

"It's not too bad," Brian laughed. "I never thought I'd be so busy. This is my first real day out since I started."

"Don't talk to me!" Larry sighed. "Father Collins thinks I'm visiting my sick aunt in Drogheda. I did visit her of course – for about five minutes. I tell you, I cut some corners on the way down here."

"You have a car?"

"Of course. A country curate has to have a car."

"If he can afford it."

"I had a bit of luck with the horses. I had my first wedding a month ago. It happened to coincide with Navan Races." He winked at Sheila. "A good time was had by all. Now are we going to this match or are we not?"

They joined the excited throng, which moved in little knots along the road to Croke Park. Rival supporters exchanged banter – friendly for the most part – along the way. Larry stopped to buy a huge rosette which he pinned proudly to his lapel.

"Savin' your presence, Father," a red-faced Cavanman taunted him, "but you'd be better off sayin' a Mass for them fellas of yours."

From the moment the match got under way Larry was a man possessed, urging on his side, finding fault with the referee's decisions and criticising the opposition's moves. Brian and Sheila found his performance more entertaining than the game itself.

"Ah, sweet ball, Frankie, sweet ball!" "Leave it to Paddy Hands?" "Ah referee! Wake up, will you?" "Couldn't kick snow off a rope!" "The man in the cap! Me life on you!"

260

When the final whistle blew and Meath were victorious by four points there was no containing Larry's emotion. He did a little dance on the terrace, got involved in animated conversation with fellow countymen and eventually made his way onto the pitch for the presentation of the cup.

Brian and Sheila remained in their seats.

"Do you think he'll forget all about us?" Sheila asked.

"It's possible. I've never seen him so excited."

"Stevie would have enjoyed all of this."

"I think of him during every Mass I say."

"That's nice." She turned to him, her broad smile fading to a serious mien.

"Are you happy, Brian?"

The question startled him. A bunch of happy Meath supporters barged past.

"I suppose I am. It's not a question I ask myself too often! Why do you ask?"

"Don't mind me!" she laughed. "I ask too many questions. My mother always said that."

"Speaking of which – how is she?"

"Not great. She has grown feeble over the last few months."

"I'm sorry to hear that."

To their surprise, Larry made his way back to them. He flopped into the seat beside Sheila, breathless from his celebration.

"Well," he panted, "what a great and glorious day! *De profundis Midensis clamavi ad te, Dominum! Gratias tibi, Domine!*" He held his arms up in grateful acknowledgement.

"I suppose you'll be going out on the town tonight," Sheila jibed.

"Indeed not. Cripes!" He looked at his watch and jumped to his feet. "We'd better move – at least I had better move!"

"What's wrong?" Brian asked as they moved to the exit.

"I have a sodality and rosary at seven. Father Collins did it on purpose, I'm sure. Can you imagine the turnout I'll get tonight? They'll all be out lighting bonfires, bar a few Holy Joes and Josephines – "

They jostled their way through a surging crowd. Brian had to lean across Sheila's shoulder and press against a wall to protect her from the rush. Her hair brushed his forehead and for a few minutes her breath was warm and delicious on his face . . .

Larry took his leave of them at the North Circular Road.

"My carriage awaits – as does my congregation of ten! It was great to see you again – both of you!"

"Thank you for thinking of us," Brian said. They shook hands warmly.

"Ah sure we'll do it all over again next year. Same time – same place?"

"Same teams?" Sheila laughed.

"Well our lads will be here anyway. I have to run. Good luck. Up Meath!" He skipped away down the street from them.

"I have a train to catch at seven," Brian said.

"I'll walk with you to the station – if you want."

"That would be nice. I might get lost otherwise," he laughed.

They reached Talbot Street as the Angelus bell rang out from the Pro-Cathedral.

"I still have an hour. Would you like a cup of tea, something to eat?" he asked somewhat nervously.

Sheila nodded. "A cup of tea would be just grand." They found a small restaurant near the railway station and ordered tea and cakes.

"Will you not have something more substantial?" Sheila asked.

"No. Lizzie will have a big tea ready for me."

"Lizzie?"

"Our housekeeper. She's a wonderful cook."

"She has you spoiled, obviously!"

They were at ease in each other's company, allowing the conversation to drift and meander at will. The hour passed all too soon. They parted on the steps of Amiens Street station.

"I really enjoyed the day," Brian said, taking her hand. "Thank you."

"Thank Larry – and Stevie!" Sheila said with that broad smile that reminded Brian so much of her brother.

"Johnston, Rooney and O'Brien," Brian laughed. "I hope we will – keep in touch."

"I hope so too. Now run or you'll miss your train."

She stood on the lower steps, waving to him until he disappeared into the station.

Brian was lucky to get a window seat on the train. There was an air of gloom in the crowded carriage which was for the most part full of dispirited Cavan followers. I must be the only happy person here, Brian thought. Happy. Are you happy? she had asked. The train jerked into movement and rolled through swirling steam out of the station. What is happiness? he mused. What is truth? Questions. He remembered another question from the seminary retreat. Are you prepared to pay the price?

CHAPTER THIRTY-FIVE

"THEY ARE BROTHER AND SISTER," ROSE SAID QUIETLY. "IT'S ONLY RIGHT they should meet – "

"Not in this house," Winston Pollock muttered. "I'll not have a priest of Rome in this – "

"Do I need to remind you that it's my house?" Rose was conscious of the tremor in her voice.

Winston flashed a look of total disdain at her. "We are man and wife, woman." He strode across the kitchen towards Rose. Her body went rigid. He stood directly in front of her, his eyebrows arched in that look that she dreaded so much. "If a priest crosses this threshold, I'll not be here to greet him. That's all I'm saying – for now!" Two giant strides took him to the door which he slammed behind him, causing the cups to jingle on the dresser beside the trembling Rose.

Brian knocked at the door of Sloe Hill with some trepidation. He looked forward to meeting his sister but this house had always made him uneasy. He gathered the collar of his coat about him. It was a week before Christmas. A biting east wind needled his face. The door opened.

"Brian! Come in. You're welcome. You must be petrified!"

"Thank you, Aunt Rose." He was taken aback by her appearance. Her hair had gone totally grey since he had last seen her but it was the haunted, drawn look on her face that jolted Brian. That and the discoloration about her left eye. "It's – not a day for – standing about," he stammered.

"How did you get here?" Rose asked, taking his overcoat.

"I got a lift – to the bottom of Caraher's Brae. I'll be picked up there again at seven."

"And you walked up the Brae? It's no wonder you're frozen. Come in to the fire." She led him into the parlour, a warm, comfortable room lit by the glow of a huge fire of crackling, hissing logs. A figure rose from an armchair – a slight man, balding and wearing clerical grey. Rose made the introductions.

"Brian, this is William Barden, Hannah's husband. William, this is Brian."

The two men shook hands tentatively and made comments about the weather and the welcoming fire.

"Hello, Brian!" Hannah stood in the doorway holding a child in one arm as she offered the other arm in greeting. She was no longer the child-sister that Brian had known. She had a mature motherly look about her. At first glance she seemed to have put on a lot of weight but as she stepped into the light Brian realised she was heavily pregnant.

"Hannah! It's good to see you." He took her hand in both of his. The child in her arms turned away at Brian's approach.

"And this is – ?"

"Moira."

"That's a lovely name. Hello, Moira!"

The child burrowed her head into her mother's shoulder.

"She's inclined to make a bit strange," Hannah explained. "And she's ready for a wee sleep, aren't you?"

"Well, motherhood suits you. You look very well."

"Thank you."

"And the second – ?"

"February."

Rose came in with a glass of cordial for Brian. "Let ye sit over to the table. The meal is ready."

They seated themselves while Rose busied herself serving the meal.

"I was saying to Hannah that Winston'll not be joining us," she explained. "There's a new batch of 'B' Specials to be trained, so he's away a lot these times."

"Doing important work for Ulster," William added. "Mr Costello and his friends are a threat to this state with their talk of reunification. Do you not think so, Brian?"

265

"I'm afraid I take little interest in politics – "

"Come now. A Roman Catholic clergyman with no interest in politics! That's not what we hear up here, is it, Hannah?"

"No. Didn't you have a chapel-gate collection for the Anti-Partition League this year?" Hannah asked, smiling at her husband seated beside her.

"I was still in the seminary then," Brian said quietly.

"Still – it was done," William countered. "But we gave those anti-partition fellows their answer in the election. Sir Basil was right. Ulster is not for sale."

Rose entered with a trayful of plates.

"Now – a grand bit of chicken and ham," she announced proudly.

Brian was glad of the interruption.

"You've gone to a lot of trouble, Aunt Rose," he said cheerily. "What happened your eye?"

The tray clattered onto the table.

"Ooops!" Rose laughed nervously and then regained her composure as she distributed the plates. "Mind now – they're hot! Och, 'twas a simple thing. A flick of a cow's tail during milking. 'Twas right sore for a while but it's grand now. And how is Bridie?"

"She's grand, fine. Enjoying the bit of farming again!"

"What happened the shop?" Hannah asked.

"She sold that – years ago," Brian replied.

"She was right," Rose said as she seated herself. "The shop was too much for her – on her own."

"William will say grace now," Hannah announced.

Throughout the meal the conversation drifted along desultorily. For his part, Brian tried to confine it to trivial matters. During the occasional lull he would steal a glance at his sister across the table from him. She had changed in more than physical appearance. She was obviously very close to her husband but it was very much a deferential closeness. She agreed with all his comments in an obsequious manner and sought his approval for her own remarks with an "Isn't that so, dear?" or "Am I right, dear?"

He wondered if things would be very different if Hannah had come on her own. He felt alone.

"I was saying – how do you like your parish?" William's voice broke into his reverie.

266

"Sorry. I was lost in thought there. It's – grand. I'm kept busy. They're football mad down there – it's like another religion," he laughed. "And what about – Castlestewart, is it?"

William nodded. "It's a fine parish. They're a God-fearing and loyal people."

"William has started a chess club," Hannah added.

"Do you play?" William asked.

"Chess? Afraid not! I never got beyond draughts."

They complimented Rose on a lovely meal and sat around the fire. A chorus of chiming clocks announced five o'clock.

"Would you like to read now, dear?" Hannah enquired of her husband. "We like to read Scripture on the Sabbath," she explained to Brian.

"You don't mind?" William asked as he took the Bible from the mantelpiece.

"Not at all," Brian said. It would be a respite from strained conversation.

As William began reading Brian relaxed and fixed his gaze on the flames that danced and weaved above the beech log . . .

For the first time in his life he was acutely conscious of division – in his own life, in his family, in the island he lived on. He felt distant from his sister. They had come from the same womb, and yet there was a gulf between them. His brother-in-law and he ministered in the name of the same God, yet he felt no fraternal feeling for William. He was in his mother's house, a few miles from where she had grown up and yet he was in another country. Hannah and William had each other. He was alone. Are you prepared to pay the price? . . .

The reading ended and the four sat in reflective mood until the chimes began again. Six o'clock.

"I must be off," Brian announced suddenly. "I have a lift to catch."

"But you've plenty of time," Rose said in surprise. "William will drive you down – "

"Not at all. I need a walk after that big dinner! And it's a grand crisp bright evening!"

He thanked Rose for her hospitality and wished all a happy Christmas before stepping out into the night. As he walked through the gates of Sloe Hill the Angelus bell rang through the clear air from Cullyboe. He blessed himself and set off down Caraher's Brae.

When he retired to his room that night, he took a writing pad and thought again of the afternoon's events before he began to write.

Dear Sheila,

I hope this reaches you before you go home for Christmas. I feel I need to talk to someone and there is nobody around here at the moment so you'll have to bear with me.

I went back to Sloe Hill (where my mother lived) today to meet Hannah and her husband. It was Aunt Rose's idea but I'm afraid it did not work out too well. Hannah was almost a stranger to me. She seems to be very much under William's influence. Although we are both men of God we had little in common. He is very suspicious of anyone or anything to do with the Free State. I just felt very much an outsider there. It's hard to feel an outsider among your own. I felt sorry for poor Aunt Rose. She did her best.

I hope we can meet some time soon – in the New Year maybe. It would be nice to talk with you. I hope you have an enjoyable Christmas and that your mother is in better form.

Regards,

Brian.

He sealed the letter and retired to bed, but sleep would not come to him. He decided to try Aunt Bridie's remedy – a cup of hot milk. It would also ease the chill that had seeped into his body while waiting an hour for his lift at the foot of Caraher's Brae. On his way to the kitchen he noticed a shaft of light from under the door of Father Maguire's room. He paused and thought of enquiring if the parish priest would like a cup of tea. From within the room came a droning sound which on closer listening emerged as a tuneless and slurred rendering of "Slieve Gallion Braes."

The reply to his letter came sooner than he had expected, in the first week of the New Year. It was the latter part of January before Brian could manage a trip to Dublin. Sheila met him at Amiens Street station. They enjoyed a long lunch in the Carlton. Brian felt a great sense of release to be able to talk about his anxieties, his fears.

"Am I talking too much?" he asked, feeling suddenly self-conscious.

"No." Sheila laughed. "I spend the whole week talking, shouting even. It's a welcome change just to listen."

The afternoon was bright and crisp. They decided to stroll along the north quays as far as Queen Street Bridge where they watched a pallid sun hover above the upper reaches of the Liffey before they turned down the south quays and back to the railway station.

Five hours of escape. Five hours of renewal. It was an interlude that would be repeated at varying intervals over subsequent years.

Minnie McCabe was dying. It did not surprise Bridie to hear that her son Jimmy was on his way home from America to see his ailing mother.

"She'll hang on now for Jimmy, nothing surer," Josie said when Minnie was eventually moved to the County Hospital. And so it happened.

Bridie was apprehensive about Jimmy's return home. They had not corresponded since his previous visit when he had rescued her from litigation. The discovery that he had rescued her with his cheque-book rather than his "superior legal knowledge" had both angered and embarrassed Bridie.

Once again he caught her by surprise. She was leaning over a gate feeding two calves from buckets when the car pulled up.

"Any room at the trough?" he called. Bridie recognised the voice but did not look up.

"If you can fit your head in a bucket," she said.

"Jeez, Bridie McKevitt, but you're still a hard woman!"

"Life makes you hard, I suppose." She turned to face him and was taken aback at what she saw. His casual clothes hung loosely from his spare frame. Deep-sunk eyes accentuated the gauntness of his face and above them a greying hairline had receded considerably.

"You look great, Bridie. The years have been kind to you."

She wiped her hands across the fold-over apron.

"I'm covered in calf-meal. I'll not shake hands, in case – "

He took her right hand warmly in his hands.

"Of course you will. What's a bit of calf-meal?"

"How is your mother?"

"She's poorly, Bridie. Poorly." The calves began pucking the buckets against the gate.

"Shoo! Enough now! Enough!" Bridie waved her arms at the calves and retrieved the buckets.

"I called in to Podgy's, only to be told you were up here."

"I'm here now the best part of ten years."

"You never wrote – "

"No." The calves returned to the gate, trying to force their heads through the bars.

"You're busy, Bridie, and I have a few things to do. I want you to do something and I won't take no for an answer. I want you to come for a drive with me to Dundalk this evening and we'll have a bite to eat."

"What would your wife have to say about that – drivin' around with another woman?"

"My wife doesn't enter into it, Bridie."

"Of course she – "

"Mary Kate died three years ago."

The waitress took their order and left.

"You accuse me of not writing," Bridie said. "Why didn't you write yourself?"

"There never seemed to be the right time. Mary Kate had a slow and painful death and when she died I was left with three girls to care for. I did write home."

"We were never told." Bridie fingered the knife nervously. "About what you did, the last time – "

"That's all in the past, Bridie."

"It may be, but it shamed me then – and still does."

"Shamed you?"

"Yes. When I showed you the solicitor's letter I wasn't asking for charity."

"You weren't getting charity, Bridie. You were getting help – from a friend."

"But you tricked me, talking of law school and all."

"Well if I told you the truth, dammit, you wouldn't have accepted the money. And anyway I did go to law school – for three months!"

A faint smile crossed Bridie's face.

270

"That's better," Jimmy said. "Now look – "

"Steak and chips for two," the waitress announced. "Do youse want the tea now or later?"

"Later," Jimmy growled. "I hope she never comes looking for a job in Jimmy Mac's," he whispered as she departed.

"How is Jimmy Mac's doing?"

"Great. We've got fifteen branches now. 'Feed 'em good! Feed 'em fast! Feed 'em cheap!' That's our motto."

They began to eat.

"Bridie, it's not too late to start again. Would you consider coming out to the States?"

Bridie paused in the middle of cutting a piece of steak.

"Is this one of your jokes?"

"It most certainly is not. You'd have a real fine life over there – and the girls would love you. I know they would."

"Are you asking me to marry you?"

"Well now that you mention it – yes!"

Bridie put down the knife and fork quietly.

"It's a bit late in the day for that, don't you think?"

"No I don't – else I wouldn't be asking you!"

"Ach, Jimmy. It's not a young *girse* you're asking."

"I'm no spring chicken myself."

"No."

"Look, Bridie, you don't have to – "

"The answer is no, Jimmy."

He put down his knife and fork.

"Don't suppose you'll tell me why?"

"I could give you a hundred reasons. I'm set in my ways. I belong here, not in the States. I couldn't stick the way of life over there."

"You could give it a try – "

"My place is here. It's the only life I know – and want to know. If I can raise a few calves and sell a few eggs, I'm happy."

"Still Queen of the Cowshed?" he said with a resigned laugh. "Would you not just come over for a holiday? A few weeks? You deserve that."

"I'd be like a fish out of water."

He looked away and sighed.

"I'm sorry, Jimmy. I really am."

271

"Me too, Bridie. Me too."

They ate in silence before Jimmy enquired about Bridie's family and especially Brian.

"I hope he'll come to Mother's funeral," Jimmy said.

"You'll need him," Bridie replied. "The Canon has grown very feeble."

"The Canon? He's still alive?"

"He was ninety a few weeks back."

"Christ, Bridie, he was ninety when we were young."

Bridie smiled. Life had not completely extinguished Jimmy McCabe's humour. The waitress returned noisily with a tea-pot, cups and saucers.

"Will youse be having a sweet?"

Jimmy looked straight into her sullen face.

"We sure will, honey – but not in this soda parlour."

A week later, the day after his mother's funeral, Jimmy McCabe left Cullyboe. In the cemetery he appealed to Bridie to change her mind but she was adamant. He shrugged his shoulders, his slight frame looking spent and fragile.

"Well, if you change your mind, you know where I am."

Brian came in on the end of the conversation.

"What was that all about, Aunt Bridie?"

She looked away from him.

"Ach, nothing. Something that should have happened many years ago."

CHAPTER THIRTY-SIX

THE MISSIONER PROPPED THE PICTURE OF THE VIRGIN MARY ON THE PULPIT ledge.

"And I repeat to you, my dear people of Drumcash, the words of our Saviour on the cross when He said to the beloved disciple 'Behold thy mother.' Behold your mother, my dear people. Have recourse to her in your daily lives. She is ever present in our lives especially in this Marian Year. She is the seat of wisdom, the refuge of sinners – "

"Pray for us," Katie McShane intoned loudly from beneath the pulpit. There were embarrassed titters in her immediate vicinity. The missioner seized his cue.

"Pray for us indeed, O Holy Mother of God. For Mary is our friend, our ally in time of need. You all know the lovely story of Joe." He paused, and allowed the barrage of coughing to subside. "Joe was no saint, but he did many good things in his life. And it happened that he died. It will come to us all, my dear people. That is the great certainty. So Joe died and he went to the gates of heaven. St Peter and Jesus looked up Joe's record and sadly shook their heads.

"'I'm sorry, Joe, but your record suggests that you need a spell down below in Purgatory,' Peter said.

"So poor Joe trudged off, downcast, regretting all the wrong he had done during his life on earth.

"Later that day Jesus was strolling through the gardens of Paradise when who does he meet but the bould Joe, with a smile on him from ear to ear.

"'Hold on there,' Jesus said. 'I thought we sent you to Purgatory.'

"'You did indeed, Lord,' said Joe, 'but as a last chance I went around to the back door, had a word with your mother – and she let me in!'"

A ripple of laughter ran through the congregation.

"So turn to Mary, my dear people. Mary Queen of Ireland – "

"Pray for us!" the voice responded from beneath him.

"And we turn to her now with her favourite prayer, the rosary."

A week later Brian dallied in the church grounds before hearing evening confessions. It was a warm sun-drenched May evening. He envied Father Maguire who had slipped away earlier to go fishing on Lough Gash. If confessions did not detain him too long this evening he would try to make a visit to Drominarney where Uncle James's health was giving cause for concern. He recited the Angelus in the shade of the yew grove before entering the church. Surely after a week-long mission there would be little demand for confession. He sighed resignedly on seeing four full pews of penitents awaiting him. He had just settled into the confessional when he became aware of a commotion outside. He threw back the curtain. Katie McShane was bending over a figure that knelt with head deeply bowed before the shrine of the Mother of Perpetual Succour. The figure, obscured by Katie, was praying fervently and gave occasional loud agitated moans. Brian made his way towards her. Katie turned to him.

"It's Susie Carron, Father. She says she's seen the Virgin Mary."

When Brian had despatched the penitents in record time, he marched briskly back to the parochial house where he had left Susie Carron in Lizzie's care. Susie sat perched on the edge of a sofa in the parlour, a cup of tea and two biscuits untouched on the table before her. She was a slight girl whose sallow complexion and cropped hair made her look even younger than her seventeen years. She clutched a rosary beads between joined hands.

"Now, Susie," Brian began, unsure of how he should handle this matter. "Would you like to tell me what exactly happened?"

"I seen her, Father. It's the honest-to-God truth." Her eyes stared wildly up at him.

"I know, Susie. I know. But just tell me what you saw – and heard. Where were you, first of all?"

274

"I was bringin' a can of milk to Granny and I took the short-cut over by Tobermurray – "

"You were by yourself?"

"Yes. And I never pass the holy well without saying a prayer. So I knelt down and I said three Hail Marys and blessed myself with the water from the well. Then all of a sudden, I got a kind of warm feeling and I looked up and there she was."

"Who?"

"Our Lady. She seemed to be standing on top of the hawthorn bush at the back of the well. There was a fierce bright light all around her."

"How did you know who it was, Susie?"

"She was just like she is in the church. She was all dressed in blue and she was smiling. She was real beautiful, so she was."

"And did she – say anything to you?"

Susie shook her head.

"But she had the rosary in her hands and she kept on smiling and nodding. I think she was telling me to say the rosary."

"And how long did she stay there?"

"I don't know, Father. A few minutes at least."

"Did you speak to her?"

"No. I was – afraid to."

"So what did you do – when she disappeared?"

"I ran back here. I got such a fright. Janey, I just remembered – I left Granny's milk at the well."

Father Maguire paced up and down the kitchen of the parochial house.

"O Lord preserve us from the Carrons – and his Blessed Mother preserve us too. That Maisie Carron has a lot to answer for."

Brian poured him a cup of tea.

"Why so?"

"She's a fanatic. Of the worst kind. That child, Susie, went off to be a nun two years ago. Her mother claimed it was a result of conversations she had with the Holy Ghost. Susie was home again in six months. And now this!" He clattered the cup and saucer down on the kitchen table.

"She was very agitated. She did see something."

Father Maguire looked askance at his curate.

"Don't tell me you believe her?"

"I – I just wonder what she did see that upset her?"

"See? See? She didn't see anything. It was just – imagination, a trick of the light, anything. You'll have to play this down, Brian. Otherwise we'll have a major problem."

He lit a match and drew repeatedly on his pipe.

"Hopefully it will blow over. Although I doubt it. Once Katie McShane got word of it, the whole parish will know about it by tomorrow evening."

Father Maguire's worst fears were realised in the days that followed. On the very next evening Maisie Carron brought Susie back to the holy well at Tobermurray. A small group of curious neighbours accompanied them. Maisie led the group in prayers and watched her daughter intently. They had been praying for an hour when Susie's face suddenly lit up as she fixed her gaze on the hawthorn bush. Maisie held her hand up to halt the prayers.

"She's here!" Maisie whispered aloud. Susie stood up, smiling broadly and nodding occasionally. The hushed group watched in awe for a full ten minutes until Susie turned towards them still smiling and said quietly.

"Pray the rosary for all the evil that's in the world. That's her message."

The onlookers immediately fell to their knees while Maisie intoned the opening decade of the rosary.

As news of the apparition spread, the crowds grew. Within a week hundreds converged nightly on Tobermurray – a normally desolate spot which was only accessible on foot through a number of fields. The unprecedented traffic on the narrow approach roads caused headaches for Sergeant Bradley who was summoned each night to put some order on the chaos that followed each gathering. Equally, Sam Kellett, through whose land the crowds trudged each evening, was distinctly unhappy. He put up a sign proclaiming the land to be private property. It was demolished on the same evening.

"Bloody oul' Protestant begrudger."

"No respect for another creed," the worshippers said.

Finding he could not stem the tide, Sam put a proposal to the sergeant which he said would be to their mutual advantage. He opened up two of his fields as carparks, charging one shilling admission "to compensate for damage and inconvenience."

Sergeant Bradley banned parking on the road "to avoid congestion." The motorists had no option but to pay their shillings to Sam and his son Alec who each manned a gate to their lands.

"The manest trick ever! Makin' money out of people's religion."

"I'd say his ancestors were in the temple when Christ threw them out for lendin' money!"

The Kelletts were not the only ones to capitalise on the apparition. Two van owners set up religious goods stalls in the field next to the holy well, having negotiated a special parking fee of a half-crown with the Kelletts. The crowds continued to swell, particularly when first the local and then the national newspapers reported the visions experienced by Susie Carron.

Brian felt bewildered by the pace of events. Both he and Father Maguire spoke cautiously at Mass about the apparitions, appreciating the value of prayer but stressing the need for a well-grounded faith. A postcard arrived for Brian.

First Lourdes, then Fatima, now Drumcash. Have you started building the hotels yet?

Larry R.

Father Maguire was becoming increasingly distressed by the way his admonitions were being flouted by his flock. He eventually appealed to his bishop who despatched a monsignor to investigate the apparitions. The monsignor advised the local priests to say and do little about the events that had brought national attention to Drumcash.

"It is our experience that these things fizzle out in their own time," he said. "It's best to distance yourselves, stay in the background."

Father Maguire took his advice literally. He retired to his room early in the summer evenings. Brian had by now become accustomed to the long, often animated monologue which usually culminated in the tuneless rendition of "Slieve Gallion Braes". Within the four walls of the parochial house, there was an unspoken acceptance of Father Maguire's condition.

The monsignor's advice enabled Brian to make regular visits to Drominarney, where James McKevitt's health had deteriorated considerably. His breathing was laboured and obviously painful, but he steadfastly refused to go to hospital or even to give up work.

"I had enough of hospital in my youth. It's not a place I want to be." He lit a cigarette amid a severe bout of coughing.

"I doubt if the cigarettes are helping you," Brian said.

"The cigarettes have been my friends all my life. I'll not abandon them now. I'll not be a burden on anyone. And as for the saddlery, sure it's the only thing I know. I have to do it."

The monsignor's forecast regarding the apparitions proved accurate. Within a month the novelty had worn off and the crowds began to fall away. Father Maguire observed over breakfast one morning that the loss of interest in the apparition was not unrelated to the opening round of the county football championships. In August, Susie Carron left Drumcash for a convent in southern England. Her mother explained that Susie wanted to continue her life of prayer there, but, for the rest of her working life, Susie's life would be one of heavy domestic toil in the remote convent.

In late August another card arrived from Larry Rooney.

The boys have done it again. Last Sunday in September. Same time, same place, same trio???

Brian smiled. He would let Larry organise the "trio". He had not seen Sheila since early June as she had spent the summer holidays at home in Cork.

When they met outside Barry's Hotel on All-Ireland Final Day, Sheila gave no indication to Larry of her friendship with Brian. It was a superb performance, Brian thought. He conveyed this to her with a knowing smile while Larry's attention was distracted. Sheila slipped into a sweet shop on the way to Croke Park.

"Well old pal," Larry began cheerily, "five years chalked up! Tell me, are you still entertaining doubts? Speaking of which – " he nodded in the direction of the shop " – isn't 'doubts' looking terrific?" The irony of his questions almost caught Brian off guard but he asked about his parish.

"Oh we're flyin' like Frankie Byrne." Larry laughed. "Father Collins has been despatched to a boarding school – I pity the poor divils there! – and we have a new PP – Father Fay – who is a gentleman, a football fan and a racing man – in that precise order!"

"You certainly landed on your feet there," Brian said.

"The answer to prayer – *De profundis Midensis clamavi* – remember!"

He moved closer to Brian. "Between you and me, I own a quarter-share in a horse. He's called The Man in the Cap – after the great Peter McDermott. When he's right – " Larry winked " – I'll let you know!"

Sheila emerged from the shop. Larry stood in admiration of her.

"Isn't it just as well Meath only win the All-Ireland every five years or I'd have lost my vocation long ago?" Sheila blushed and made a playful gesture of disapproval towards Larry.

Once again, as soon as the game began, Larry was transformed into a boisterous argumentative fan and when Meath emerged victorious over Kerry he was delirious with excitement. Brian and Sheila waited for him to emerge from the exultant crowd but he never appeared.

"Easy knowing there's no sodality this evening!" Brian laughed.

They made their way to a restaurant in Lower Abbey Street where Brian realised how much he had missed Sheila as a confidant over the summer. She listened with awe and delight to Brian's account of the Marian apparitions. As Brian was concluding his story, a tall figure passed his table, paused, nodded and moved on. It took a while for Brian to recognise him as the monsignor who had investigated the apparitions.

On the morning of Hallowe'en, 1954, a telegram was delivered to Brian.

James very ill. Come quickly. Bridie.

Brian was in Drominarney by mid-afternoon. Bridie greeted him at the door, her eyes weary and reddened. James lay in bed in some distress. He was fighting for each breath but he brightened considerably on seeing Brian.

"Is it – time yet – for Barney?"

"Barney?"

"Barney – McCool – of Coolaghy." The voice was growing more hoarse but Brian recognised James's favourite programme on the wireless.

"Put Barney – on – there – Brian – Barney's – the wild – man – for the stories." His attempted laugh only worsened his breathing.

Brian sat by the bed and took out *The Garden of the Soul*. He ran his thumb slowly over the worn leather cover and remembered a day's shopping in Castleblayney when James had bought this book for him. He opened it at the "Recommendation of a Departing Soul."

"May the bright company of angels meet thee; may the court of Apostles greet thee – "

"Tell – Gussie Fox – I'll have – the blinkers – ready for – him – tomorrow – "

"And like as wax melteth before the fire, so let the wicked perish at the presence of God . . . "

"Pete – would want – to be puttin' – bluestone – on the spuds – in this heavy – oul' weather – "

"May Jesus Christ, the Son of the Living God, place thee within the ever-verdant gardens of Paradise . . . "

"I saw – Sarah – goin' up – the brae – D'ye mind – the time – she cut – her head – rollin' – down – "

Bridie, holding James's hand, uttered an anguished sob.

"Mayest thou enjoy the sweetness of the contemplation of God for ever and ever. Amen."

James McKevitt was laid to rest beside his brother Pete on the feast of All Souls, 1954. Later that evening Brian climbed the stairs to the loft for the last time. He stood at the workbench, fingering off-cuts of leather and inhaling their unique smell. The wee hammer hung on a nail by the window. He remembered the many evenings when James would let him play with the hammer, when Brian's only ambition was to be a saddler like Uncle James. He slipped the hammer into his pocket as a memento of the man he loved as a father.

Two months later he received a letter from a firm of solicitors in Castleblayney. Its message was brief.

"Please find enclosed a cheque for £828, being part of the estate of James McKevitt, Drominarney, willed to you by the aforesaid, on the stipulation that you offer a novena of Masses for his soul."

Brian invested the money in the purchase of a Ford Prefect car which greatly facilitated his pastoral work particularly during the harsh winter months. It also afforded him the opportunity to make more regular visits to Dublin.

CHAPTER THIRTY-SEVEN

F ATHER MAGUIRE'S HEAD TILTED SLOWLY BACKWARDS OVER THE TOP OF THE car seat. He began to snore. Brian was initially startled from his concentration on driving, but then found his companion's comical pose and sporadic snoring highly amusing. They were returning from the funeral of Canon McKenna of Cullyboe. Twenty-eight diocesan priests had joined in the Solemn High Mass at the funeral. They had been entertained afterwards in the parochial house. Brian noticed how jovial Father Maguire had been among his peers – chatting and joking with men he rarely met in such numbers. A few glasses of whiskey added to his conviviality. He slept through most of the journey home.

As they approached Drumcash Brian deliberately braked sharply, jolting the parish priest awake.

"Wha – ? Oh, must have dozed off! Tiring affairs, these High Masses."

Brian hummed in agreement. Father Maguire retired to his room earlier than usual that night. Brian relaxed in the parlour with the newspaper and the wireless. He flicked the tuning dial of the wireless slowly clockwise until he recognised a fragment of a Mozart piano concerto. He thought of the "soirées" in Father Vincent's room in St Felim's many years ago. Father Vincent. Brian wondered if he was still in England, still alive even. He glanced at the newspaper headlines – a deepening crisis in Suez, an uprising savagely quelled in Hungary and, much nearer home, the closing of several border roads following increased IRA activity. Conflict and separation.

Caraher's Brae had been cratered as he had discovered when attempting to pay a visit to Sloe Hill . . .

He had left his car on the southern side of the border and walked to the house. When he knocked at the front door he was startled to be confronted by Winston Pollock caressing the barrel of a rifle.

"You can't be too careful in the times that are in it," he said.

"I suppose not," Brian agreed.

"You'll be looking for the women and the Reverend? They're away in the five-acre field working at the hay."

Rose, Hannah and William Barden and their children, Moira and Roy, were busy turning the hay.

"We could do with an extra hand," Rose jibed on Brian's arrival.

"Depends on the wages." Brian laughed.

"There's a mug of tea and a heel of good soda bread. I'm away now to make the tea."

"Not at all," William interjected, handing a fork to Brian. "The children are getting tired. I'll take them to make the tea."

It was good to feel the smooth handle of the hayfork between his hands again and to inhale the rich scent of the hay as he worked his way steadily along the row, parallel to his sister and his aunt. Hannah was less guarded than usual in the desultory conversation of work. If Brian closed his eyes he might as easily be in the company of Hughie and Bridie. The laughter of the returning children broke the spell. The group sat at the headland relishing the hot tea and soda bread dripping with raspberry jam. The conversation took a definite turn.

"I see where one of those IRA boys blew himself to pieces when preparing a bomb," William Barden said as he watched his children leap the hay-rows. "The wages of sin is indeed death."

Brian picked his way gingerly around the craters on Caraher's Brae and made his way back to the car. Conflict and separation . . .

"You're – still here! Good!" Father Maguire's voice startled him from his reverie. It was most unusual for the parish priest to reappear downstairs once he had retired for the night. He stood unsteadily at the door with a glass of whiskey in his hand. His shirt was partly unbuttoned and his braces hung by his side. He was aware that his dishevelled appearance was causing Brian some embarrassment.

"You – will pardon – my appearance," he said as he tottered towards his armchair. Brian rose to assist him but he waved him away.

"I have something that needs saying – and it can't be put off any longer." He eased himself into the chair and took a long drink before continuing.

"It has been brought to my attention – as they say – that you have been seen in the company of a woman – in Dublin." The unexpectedness of the words sent a chill through Brian's body.

"I have no wish to know the details but suffice it to say those in authority insist that you terminate this – relationship – at once."

The word "relationship" stung Brian from his silence.

"But there's no relationship. There's just friendship – company."

"It matters not what you call it. It is not approved of by those in authority."

Those in authority. Who? Who could have known? In an instant he had a fleeting memory of a monsignor nodding in recognition in a Dublin café.

"It's a small country, Brian. No more than a big parish. Nothing goes unnoticed."

Brian sighed in frustration.

"She's just the sister of a friend. A very dear deceased friend. She's just company. There's nothing – "

"The priesthood is a lonely job. I know that, Brian. We all seek company in our own way." He held the glass up to the light to confirm that it was empty. "The trouble is that in this country my company is more acceptable than yours."

Brian sank deeper into the armchair. He felt a great weight pressing in on him. Father Maguire replenished his glass from a bottle which he took from the bottom of a bookcase.

"I have done my duty in conveying the wishes of the authorities." He resumed his seat. "Now let me tell you something in confidence. I have been down your road. Forty years ago I had to make a decision. I made the wrong decision and ruined a lovely girl's life. She married within two months of my decision. Married a scoundrel who beat her every day of her life until she could take no more. She just stopped living. Her blood is on my hands." He swallowed the glass of whiskey in one gulp.

Brian's mind was in turmoil. He sat up late into the night pondering his situation. In the eyes of "those in authority" he had done wrong. In his own mind he knew that this was not so. He wondered if anything had been said to Sheila. She was at home in Cork for the summer. What would he do in September? Would he obey the instructions of those in authority? What would he say to Sheila? The question from the seminary retreat came back to haunt him once more. Are you prepared to pay the price?

A few weeks later a letter arrived. The postmark – Sussex – intrigued Brian. Who did he know in Sussex? He tore the letter open eagerly. The opening sentence answered his thoughts.

Dear Brian,

I can see you opening this letter and saying to yourself – Who do I know in Sussex? Well, I'm not sure if you know the writer. He used to be Father Larry Rooney, CC Balgower (the town of the goats!) He is now plain old Larry Rooney, head lad at The Willows Stud – and husband of Susan, the owner of the aforesaid!

Are you shocked? How do you think I feel! Rooney fell at Beecher's Brook. It was her lovely legs, compact body and the way she carried herself that did the damage. And that was only the filly that Susan was buying at the bloodstock sales!

Seriously, we met three years ago at the sales and, as they would say in Balgower, we clicked straight away. I fought the good fight for as long as I could but in the end I succumbed – and here I am in deepest Sussex for the past six months. I suppose I've committed three sins – left the priesthood, absconded to pagan England and married a horsey Protestant! I feel sorry for my poor mother, for the people of Balgower and for Father Fay – a lovely man. Maybe what I've done is mad. Maybe entering the seminary was mad. I'm a bit confused at the minute. Pray for me.

So there it is. Johnston, Rooney and O'Brien – and then there was only Johnston . . . Are you still entertaining doubts? If so, give her my regards! The priesthood is a great calling, Brian, but in the end I suppose I wasn't "prepared to pay the price." I was only an each-way bet at best!

Regards,
Larry.

The letter both saddened and troubled Brian. He felt the loss of a friend and a fellow priest but he also felt challenged by Larry's decision. The certainty of ten years ago was crumbling about him.

A huge white banner hung above the entrance to Cullyboe church, its gold lettering glinting in the September sun.

Welcome home, Father Pat

25 years of service to God

Deo Gratias

"Who's Father Pat?" Brian asked.

"Do you not remember him? I mind him giving you two half-crowns for your First Communion." Bridie struggled into the front seat of the car with a heavy basket. "Och, my old knees are not what they used to be." She sighed.

"Too much praying," Brian suggested with an impish grin.

The car moved off. Two half-crowns. He could remember them but not the donor.

"He's been away working on the missions. This twenty-five years," Bridie said, reading his mind. "Some say he's coming home to die. Seems he picked up some strange sickness in Africa. Aye!" She tried to massage some ease into the acute pain in her left knee. "Pat Trainor. He wasn't one of my favourite people."

"Why?"

"I don't know. I couldn't warm to him. He went out with your mother for a while."

"He – what?" Brian gripped the steering wheel tightly.

"Och, 'twas nothing really. She was young and he must have been ten years older than her. She went to a dance in the schoolhouse with him and after that he had a kind of hold over her. He was that kind of person. I think she was afraid of him. Then she met Gordon – "

"Where? Where did she meet him? You never told me – "

Bridie laughed. "It wasn't the most romantic of meetings. She was bringing a can of tea to the boys at the hay when her bike got a puncture – and along came Gordon. That was the start of it – and Pat Trainor didn't like it one bit. He continued to pester her but she stood up to him. Then, about the time Sarah got married, Pat surprised us all by going off to become a priest. It was a strange thing, surely."

285

Brian turned the car down the lane to Bridie's house.

"And then there was Aggie. The creature. I wonder if she's still alive."

"Who's Aggie?"

"Pat Trainor's sister. God love her – she was never right in the head – but there was no harm in her. When Pat was going away there was no one to look after her, so he had her sent to the County Home. She may be still there, for all we know."

They reached the house.

"Will you come in?" Bridie asked.

"No, thanks. I have to visit the school in Drumcash. First meeting with the confirmation class."

The telephone rang in the presbytery. Father Maguire answered it.

"It's for you, Brian," he called.

Brian took the cup of tea with him from the kitchen.

"Is that Brian?" The voice was hoarse and completely strange.

"Yes."

"This is Pat Trainor. You wouldn't remember me."

Brian was surprised into silence.

"I've just come back from – "

"I – I know. You're welcome home."

"Thank you." There was a prolonged muffled cough. "It's good to be home – for however long."

"Can I – help you?"

"I hope so. Are you – hearing confessions tonight?"

"I'll be starting in twenty minutes."

There was a long pause.

"Would you – hear mine?"

"Yours?"

"Yes. What time would you be free?" He seemed to accept that Brian would have no objection.

"I'm usually finished by eight o'clock."

"I'll be there at eight."

CHAPTER THIRTY-EIGHT

B RIAN WAS DISQUIETED BY PAT TRAINOR'S PHONECALL. WHY SHOULD HE ask for Brian as confessor? He felt a sense of foreboding as he made his way across the churchyard. He could not concentrate properly on the confessions he heard and dismissed the penitents with an unaccustomed abruptness. By seven-thirty the church was empty. He opened *The Garden of the Soul* in search of solace for his troubled mind . . .

"Thou art my sure rock of defence against all kinds of enemies, thou art my ever-present grace, able to strengthen and to conquer. In all my sufferings, therefore – "

The church door creaked open. Ten to eight. A slight figure moved through the pews. Brian's heart began to race.

" – in all my weakness, will I confidently call upon thee – " The confessional door opened. Brian slid back the shutter.

"Bless me, Father, for I have sinned – we had a cow calving, Father." It was Katie McShane. Five past eight. He wasn't coming. The interior of the church had grown quite dark. Ten past. Brian felt a surge of relief as he rose to leave. The church door creaked again. Brian sat back into his seat in the confessional. Through a chink in the curtain he could discern a slightly stooped figure making its way toward him. A laboured breathing echoed through the emptiness of the church. The confessional door opened, then closed. Brian slid back the shutter.

"Bless me, Father, for I have sinned." The breathing grew even more laboured. "I have sinned. I have sinned."

"How have you – ?"

"I have already confessed this sin – to my bishop – in Africa – but I must confess it – to you – before – before – "

"Why? Why is it important to confess it to me?" The question was almost drowned out by a rasping cough.

"Many years ago – in moments of madness that I can never explain – and that have tormented me ever since – I had carnal knowledge – of a girl – a lovely, innocent girl – against her will. It was jealousy – rage – madness."

Brian felt an acute throbbing in his temple. His throat dried up but he forced the words out.

"And the girl. What happened the – ?"

"Her name was Sarah."

The pain intensified. Brian's eyes could not focus. He gripped the ledge that supported the shutter. The voice beyond the shutter had shrunk to a whisper.

"I have reason – to believe – that that act led to the – birth – of twins."

Brian shut his eyes tightly. No. No. No. No.

"For this – and all the sins – of my past life I am – truly sorry." The voice broke into muted sobs. "Truly – sorry. I ask your forgiveness – as well as – the Lord's."

Brian suddenly felt cold. His whole body began to shudder. The more tightly he shut his eyes, the more clearly he could see a misshapen mound of earth marked by a crude lichened stone. He felt the beads of sweat forming on his brow. The leather cover of his prayer-book became clammy in his grasp.

"Pray for those – you have wronged," he whispered.

"I have done so – daily – for the past thirty-two years," came the reply.

"*Ego te absolvo . . .* " Brian could hear the words but they seemed to come from beyond rather than within him. His hand trembled as he made the sign of absolution.

"Forgive me, Brian. Forgive me." There was a long silence punctuated only by the breathing of both men. Brian willed himself to speak but no words would come. Finally the penitent rose with difficulty and left the confessional. A shuffling across the side aisle and then the door creaked shut.

Brian sat in the confessional for a full half-hour. He welcomed the gathering gloom. He willed it to envelop him, drown him, erase his identity totally, but the hands now shaking before his eyes were real flesh and blood, created by an act of savage violation. He shoved them deep into his pockets. He shook his head violently as if to dislodge the images that flitted relentlessly across his brain. Each one of them concerned his mother – a shy girl looking downwards in a faded photograph, a happy young woman swinging him round playfully in a recurring childhood dream, a body floating in a river, a rough mound of earth in a lonely field. And then the most repelling image of all . . .

He jumped up and marched briskly to his car. He paused only to exchange his jacket for an old navy jumper that lay in the back seat. He drove away at speed through a heavy drizzle. He had to get away from Drumcash. To Dublin. To Sheila. On an impulse he swung the car away from the main road and headed for Cullyboe.

He parked outside the cemetery and waited until two figures on bicycles, heads bent into rain, passed by. He hastened across the road, over the stile and through the long grass of the little field opposite the cemetery. He stumbled a few times as he tripped over hidden stones. When he reached his mother's grave, he sank to his knees in the drenching grass and cried angry tears.

"Forgive us, Mother. Forgive us all – for the way the world treated you. What a cross you carried and this is how you were judged." He swept his arm about in a wild, futile gesture. He knelt there in silent prayer, mindless of the rain which now beat down relentlessly upon him. In the gloom he could faintly discern a daisy growing on the grave. He plucked it very carefully and took it back to the car, where he slipped it between the leaves of *The Garden of the Soul.* It was only when he sat into the car that he realised how cold he felt.

The windscreen wipers laboured against the driving rain. The restricted vision, added to his distracted state of mind, caused several near misses on the road to Dublin. Other drivers sounded their horns or flashed their lights in anger but he drove on without stopping. It was approaching midnight when he knocked on the door of Sheila's basement flat in Drumcondra.

She opened the door cautiously until she recognised the dark-clad bedraggled figure outside.

"Brian! What on earth – ? Come in! Come in!"

He slipped past her into the kitchen.

"My God! You're wringing wet. Where have you been? What has happened?"

He stood in the middle of the kitchen, leaned on the table and shook his head slowly to and fro.

"Cold," he whispered. "So cold."

Sheila sprang into action. She plugged in an electric fire, filled a kettle and put it on the gas ring.

"Take off that jumper. It's doing you no good in that condition," she called as she searched anxiously in a kitchen press.

"I hope to God I have a drop of whiskey here. Aaah!" She gave a sigh of relief and held the bottle up to the light.

"Drop is the word!" She went into the bedroom and took the heaviest blanket from her bed. She draped it around Brian's shoulders as he huddled in front of the fire. The kettle began to sing. She drained the bottle of whiskey into a mug, added sugar and boiling water.

"Now – get that into you!" she said, wrapping his hands around the mug. She watched silently as he sipped awkwardly from the mug. There was a frightened look in his face that she had never seen before.

"Would you mind – turning off the light?" he croaked. She acceded to his request and sat beside him on the settee.

"Now can you tell me what's wrong?" She adopted a jocular tone. "Were confessions really that bad?" His reaction startled her. His body went rigid. In the glow of the fire she glimpsed a look of sheer horror on his face.

"I'm sorry," she whispered. "I shouldn't have said that."

He shook his head.

"I can't – explain. Would you hold me – please?"

She took his shuddering body in her arms and held him close to her. Occasionally she caressed his brow to assuage the restlessness that surged through him. No more words were spoken and as the night hours wore on, the heat from the fire eventually induced a kind of sleep in Sheila, a sleep that was broken by the violent shudders that seized Brian's body at irregular intervals.

"Will they not be wondering if you don't turn up for Mass?" Sheila asked as she prepared breakfast.

"Of course they will. They'll do more than wonder – but I just can't face back there today. I need time."

She stirred the scrambled eggs vigorously.

"And you still can't tell me – ?"

"No. I'm sorry, Sheila."

She shrugged her shoulders and spooned the scrambled egg onto the toast.

"Please understand, Sheila – "

"Breakfast is ready." She smiled.

Sunday crawled by. Sheila went out to Mass and later to the shop while Brian remained indoors.

"Of all the days to go looking for a bottle of milk and a few rashers!" Sheila sighed on her return. "I forgot it was All-Ireland Final Day. The place is crawling with Corkmen. I hope none of my relations come looking for me. This day always reminds me of Larry. Have you heard from him?"

"He's left the priesthood and married."

"Married?"

"To a horsey Protestant, as he said himself. They're living in Sussex."

"My heavens! Why didn't you tell me?"

"I didn't get a chance – until now."

"You could have written. I was very fond of Larry. You knew that – "

"Sheila, please believe me. So much has happened in the past few weeks – "

"What else have I missed?" There was a sardonic edge to her question. Brian gave a sigh of resignation.

"I have debated with myself how to tell you this – or whether to tell you at all, but now – Father Maguire informed me that it has become known to 'those in authority', as he put it, that we have been seen together in Dublin."

"And that's a big sin?"

"Big enough for them to ask that the relationship be terminated."

"The relationship?"

"That was my reaction too."

"And this is what last night was all about?"

"No. No. It had nothing to do with last night. Nothing at all. Nothing."

"How long have you known?"

"Since August."

291

She searched in the kitchen drawers until she found a cigarette pack.

"I didn't know you smoked."

"Only when I'm upset. Do you want one?"

He shook his head as Sheila lit her cigarette.

"I'm sorry, Sheila, that you've been – "

"The evil woman?" She exhaled a long column of smoke. "Isn't it little they have to be doing? They're probably outside right now. Is that the reason for this?" She gestured to the drawn curtains.

"No. I just don't like the light – at the moment."

"You can't shut it out forever, Brian."

"I know that."

She flicked the cigarette ash nervously into a saucer.

"I'd better see about something to eat," she said. "The match is on the wireless – if you're interested."

When Brian woke on the Monday morning, Sheila had already left for work. There was a note on the table.

Gone to school. Back around four – if you're still here. At least let some light in.

S.

He was still there when she got home but the curtains were drawn back.

"I took your suggestion literally – I don't know if it had a more cryptic meaning?"

Sheila shrugged her shoulders. Brian bit his lip in embarrassment. Something was happening to their friendship, their "relationship."

"I'll be going back tonight – to face the music!" he said quietly.

She dropped her bag and held out her arms to hug him. As she withdrew he could feel the moistness of her eyes on his cheek.

"I brought you a paper. You can read all about the match. It was a bad weekend for Cork." She sighed.

He waited until dark before leaving. "Thanks for putting me up – and putting up with me," he said, his hand resting on the latch.

"One was easier than the other."

"You'll keep in touch?"

Sheila nodded. "I must see about a red light to go over the door." Her voice was trembling.

"Sheila, please. Both of us know we have nothing to be ashamed of."

"Try telling your superiors that! You'd better go. Goodbye, Brian." She hugged him and turned away quickly.

Brian felt a great loneliness as he drove home to Drumcash. The sacrament of penance was meant to be a sacrament of healing and reunion. His experience of it this weekend only proved it to be one of division and hurt. He was hurting terribly himself and he knew that Sheila was, understandably, withdrawn from him. Now he had to face Father Maguire. And Pat Trainor – could he face the man who claimed parenthood of him through an act of violation?

The light was on in Father Maguire's room. Brian knocked timidly on the door.

"Come in!" The voice was gruff, impatient.

"It's me, Father."

The parish priest was seated in his armchair, an open book on his lap, a glass of whiskey on the table beside him. He did not look up.

"So. The prodigal has returned."

"I'm sorry – "

"Would they not even feed you the very husks that the swine were eating?"

"Something happened – that I can't explain – "

"Well there's no fatted calf awaiting you here, boy."

He was drunk. There was no point.

"Nothing but a disappointed congregation who turn up for Sunday Mass and there's no priest to say it. You let them down, boy. You let me down. Above all, you let yourself down. I don't want excuses. I don't care to know. I don't care . . . I'm tired. Goodnight."

Brian closed the door quietly. When he opened his own door his eyes fell on a folded slip of paper on the carpet. He picked it up.

Sat. 10 p.m.

Brian,

Nobody seems to know where you are. My brother has died suddenly RIP. I'm away home. Can you cover for me tomorrow? His funeral is on Monday. I will be back on Monday evening.

P. Maguire.

On the following Sunday Brian made a public apology at the end of Mass.

"I wish to apologise most sincerely for the situation that arose last weekend whereby you were left without a priest, particularly for Sunday Mass. I had a – personal matter that had to be dealt with urgently and unfortunately I had left before hearing of Father Maguire's sad news. I apologise to him and to you and I ask you for your prayers for him at this sad time. It was inexcusable on my part. I simply ask your forgiveness."

When he came back to the sacristy he noticed a white envelope with his name typed on it, propped against the crucifix. He opened it with some puzzlement. It was not a letter but a brief statement headed "Diocesan Appointments."

"The following appointments have been made and will take effect on November 1st 1957 . . . "

His eye ran down the list –

Fr C Brady, CC Knockmore to be CC Drumcash.

Fr B Johnston, CC Drumcash to be chaplain, St Felim's Hospital and Sisters of Mercy . . .

Brian sought out Father Maguire and found him in the parlour, leafing through a Sunday newspaper. Brian forgot the distance there had been between them for the previous week. He waved the sheet of paper before him.

"Have you seen this?" he asked.

Father Maguire peered over his spectacles.

"I have a copy."

"Is this to be my – punishment?"

"Is it punishment for the other nine priests?"

"But my move is fairly obvious – get him out of a parish where he won't be a source of scandal – "

"To my knowledge those appointments were decided before last weekend."

Brian felt totally deflated. He crumpled the sheet of paper and shoved it into his pocket. As he turned to go, Father Maguire resumed reading his newspaper.

"By the way," he said, without looking up. "Can you make sure to be here on Tuesday? I have to go to Cullyboe for a Solemn High Mass."

"Cullyboe?"

"Yes. That's your part of the world, isn't it? It's for the late Father Pat Trainor. Did you know him?"

CHAPTER THIRTY-NINE

BRIAN JOHNSTON LEFT DRUMCASH ON A SATURDAY AFTERNOON IN LATE October, with all his possessions packed into the Ford Prefect – all except one, the handsome fireside chair which the parishioners had presented to him on his departure. That would follow in a few days in Gussy McArdle's van. There had been a presentation with flowery speeches and kind words but Brian was aware that his leaving Drumcash would not be a serious blow to the community, particularly since his successor was a keen footballer. If, on top of that, he was "cushy" enough in the confessional, he was guaranteed immediate acceptance.

Father Maguire had grown more morose in the weeks prior to Brian's departure. He would retire to his room directly after the October devotions each night. Brian was often awakened by his singing in the early hours of the morning. The night of the presentation to Brian was the one exception to the parish priest's routine. On their return to the parochial house, he invited Brian to join him for a drink in the parlour.

"I meant what I said up in the hall," he said, as he sank into the great leather armchair. "You are a good priest, Brian. What I didn't say is that you have the makings of a great priest – but you also have the makings of a very unhappy priest. And you know what happens priests of that kind . . . The choice is yours. I will only say this. If the relationship is serious – get out of the priesthood now before you cause irreparable damage – to yourself and others. That's just my personal view."

The level-crossing gates were down. Brian eased the car to a halt. Serious? He smiled to himself. He could feel the crumpled envelope in his inside pocket. It had arrived a week earlier. He read it again now.

Dear Brian,

Just a short note that will hopefully find you well. I hope your absence from Drumcash had no major repercussions.

I have decided to go back to Cork. Mother is quite feeble now and needs looking after. A vacancy has arisen in a school about five miles from home. I applied and have got the job. I start on December 1st.

It will mean we will be at opposite ends of the country. Maybe that is a good thing! – I don't know. I will miss my boisterous city pupils. I will miss you too but maybe this is best.

With love,

Sheila.

Serious? It seemed to Brian that Sheila had written that letter under pressure. Maybe she had received a "diocesan appointments" letter too, he mused.

"Miss Sheila O'Brien NT, Good Counsel NS to be Assistant, Doonbarra NS . . . "

He caught sight of a familiar landmark – a line of tall yew trees behind a high wall. He slowed down as he approached a gap in the trees. A narrow drive led through the gap. A weather-worn sign stood atop the wall – St Felim's General Hospital.

The room was large, high-ceilinged and sparsely furnished. It had obviously been a ward at one time. The only concession to home comfort was the soft lemon tone in which the walls had been painted, with matching heavy drapes covering the tall, steel-framed windows. Brian's new fireside chair would be a welcome addition, even if the fire were only an electric one. The sisters had been kind and welcoming but as he sat on the edge of his iron-framed bed on his first evening in St Felim's, he could not help reflecting on his first evening in another St Felim's – the boarding school – some twenty years previously. What he felt then he felt now – a longing for the warmth and the comfort and the freedom of home. He remembered how on his first morning in the school washroom he had plunged his face into his towel and held it there because the towel had that very private and delicious smell of home. Home. He hadn't known home since then. He had been in transit, or so it seemed.

He began unpacking the cardboard box that sat at his feet. It contained mostly books but on top of the books lay an assortment of personal items. A paltry collection of trinkets but each one had a significance in his life. The wee hammer . . . Uncle James's loft had now been converted into bedrooms to cater for the needs of Hughie's growing family. A scarred and chapped leather hurling ball, on whose surface he could trace the faint initials SOB . . . "Stevie O'Brien from Cloyne, the home of hurling, boy!" A china eggcup with a rose pattern, cracked and chipped . . . he had read recently of the Reverend William Barden's call to set up a United Church of Ulster "in the face of a growing nationalist threat." Brian surveyed the collection of trinkets. Did they sum up his life to date? Loss and isolation. He picked up *The Garden of the Soul*. It fell open at a page marked by a shrivelled daisy. He picked the fragile flower up with trembling fingers. Mother. Home. Family. What if the nightmare had never happened? If he were the natural son of living parents? Would he be a priest? Or maybe working the land with his father? His father . . . He withdrew a newspaper cutting from the cover of his prayer book.

Tributes Paid to Missionary Priest. "A stalwart of the church, in the grand tradition of the early Irish missionaries" was how Monsignor James Moran, PP Cullyboe referred to the late Father Pat Trainor who died within weeks of returning home from twenty-five years' service on the African missionfields. Father Trainor, a native of Cullyboe, was interred locally after a Solemn High Mass which was officiated at by a large number of diocesan priests. "Pat Trainor's life was that to which we should all aspire – a life of service given selflessly to God and his fellow man," Monsignor Moran told the packed congregation.

Brian replaced both the cutting and the daisy. The nightmare had happened. He was a flawed person – and consequently a flawed priest.

He fell easily into the routine of hospital life. The demands of the chaplaincy were less varied than that of parish work but their intensity was often greater. Witnessing suffering each day was difficult. When that suffering was terminated by death, particularly in the case of a young person, he felt helpless. The vacant stare in the eyes of grieving parents, many of whom it was his duty to inform of the death of their child, tore him apart as he fought to find words of comfort that he knew were hopelessly inadequate. Some were less accepting than others. An anguished husband whose wife had died of cancer interrupted Brian's prayers in the mortuary.

"Compassion!" he railed. "How can you talk of a God of compassion? If He had compassion, He wouldn't have taken her. Thirty-five years of age," he sobbed. "And me left with three children under five. Where's the compassion in that?"

"I know how difficult it is for you – " Brian began.

"No, you don't know. How could you know with your nice cushy job? How could you know what it's like to hear a child screaming in the night for a mammy that will never come? Ask your God of compassion about that, will you. Go on, ask him, ask him, for Christ's sake!"

He threw himself across the rigid body of his wife, heaving great sighs from the pit of his stomach as he did so. His relatives cast apologetic glances at Brian as they sought to restrain him. Brian smiled in acknowledgement and withdrew from the room. He hastened down the corridor and crossed the covered-in veranda before making his way through the little flower-garden into the nuns' burial plot. He leaned, breathless, against the Calvary in the centre of the plot. The cold of a January evening pinched his face as he beheld the white figure of Christ, eerie in the winter light. He knocked his head in frustration against the base of the cross. He had no answers for that broken man. There were no answers.

There were lighter days and lighter times. Brian enjoyed the company in the men's wards where endless card-games and disputations about football and politics filled long days.

"Ah Christ! – sorry, Father! – where did you get that trump? You had none when I led with the trump! You're after renegin', you mean oul' get . . . "

"Clontibret have about as much chance of winnin' that match as de Valera has of gettin' a kiss from me mother . . . "

"Isn't Dev the lucky man?"

"That's not a hand of cards. It's a foot! Would you ever deal them again?"

"Aye, and give them a good shuffle like Costello did with his ministers . . . "

Brian had just completed his distribution of communion on a First Friday morning. He was invited to have a cup of tea at the nurses' station in St Joseph's ward. A young nurse took a long drink from a mug of tea.

"Oh, I needed that," she sighed. "That Trainor woman has me heart scalded."

Brian started at the mention of the name.

"I know," her older colleague replied. "The night nurse told me she nearly woke the whole ward at three o'clock in the morning chanting the litany of the rosary."

"Who's this?" Brian asked, feigning amusement.

"An old dear they sent over from the County Home for running repairs," the young nurse said impishly. "What's this they call her?"

"Mad Aggie," the older nurse replied.

"Aye, that's it. Mad Aggie. The sooner they send her back to the Home the better, before we all go mad!"

Aggie Trainor. Mad Aggie. Brian remembered Bridie's words. "She was never right in the head, the creature . . . he had her sent to the County Home."

"Where is she now?" he asked the nurses.

"Women's Gerries – sorry, Women's Geriatric – St Jude's." The young nurse giggled.

"Thanks for the tea, ladies," Brian said and moved down the corridor. The nurses nodded in surprise to each other.

St Jude's was a double room – to the left a ward of some ten or twelve beds and to the right a day-room where patients who were not bedridden could spend the day-time hours. The day-room had a grim cheerless look. Dark brown wainscot covered the lower part of the four walls. Above, a faded mustard paint flaked and peeled from the wall. A large painting of the Holy Family was the room's solitary adornment. It dominated the wall at the far end of the room. Beneath it a huge fireplace held a smouldering turf fire. A wire cage surrounded the fire. The stench of stale urine pervaded the room. There were six old women seated at intervals around the room. Their wasted bodies lay slumped and contorted in chairs that seemed to have grown too big for them. A solitary figure wearing a green coat that swept along the floor, a wisp of a woman, with thinning white hair, walked anxiously up and down the centre of the room. She was chanting in a monotone which was barely audible to Brian at first but became discernible when he entered the day-room.

"Cause of our joy – pray for us.

Spiritual vessel – pray for us.

Vessel of honour – pray for us.

Vessel of singular devotion – "

"Hello Aggie," Brian said gently.

She stared at him and turned sharply away.

"Mystical rose – pray for us.

Tower of David – pray for us."

Brian walked alongside Aggie. He decided to join in the responses to the litany and thus won her confidence.

They stood before the fire. Aggie gazed at the picture above the fireplace.

"I'm from Cullyboe, Aggie. Do you remember Cullyboe?"

She smiled and nodded her head continuously.

"I'm going to confession to Canon McKenna tonight."

Brian gripped the fireguard. "And how is – Pat?"

"Pat's away. Pat's away."

"Do you remember Sarah, Aggie? Sarah McKevitt?"

Aggie's whole body began to tremble. She seemed to wither before Brian's eyes. She blessed herself repeatedly.

"M-mother most pure," she stammered.

"Pray for us," Brian replied.

"M-mother most chaste."

"Pray for us."

"S-Sarah inviolate."

Brian bit his lip.

"S-Sarah undefiled." Aggie's voice broke into choking sobs.

Brian put his arm around her.

"It's all right, Aggie," he whispered. "It's all right. Come on over here and we'll sit you down." He guided her slowly to her chair and beckoned the nurse who had watched the entire scene, bemused.

"I think Aggie would like a sup of tea," he said.

He sat with Aggie for thirty minutes, holding the drinking cup while she drank through the spout. He held his arm around her frail shoulders until the convulsion subsided and she reverted to her original chant.

"Comfort of the afflicted – pray for us . . . "

"God bless and protect you, Aggie. You've carried a terrible burden for all those years." He eased away from her, nodded to the nurse and took a lingering look at the tiny nodding figure in the big chair before hurrying away to his room. He needed to be alone. It seemed that there was no escape from the spectres of the past.

Aggie was sent back to the County Home a week later. Brian struggled on with his ministry. It was April 1959. He had almost completed ten years in the priesthood. A milestone of some significance, he reflected. The achievement was not forgotten elsewhere. A letter arrived from Sussex, having been re-routed from Drumcash.

Dear Brian,

Ten years toiling in the Master's vineyard. Congratulations! You deserve a break. Susan and I would love to see you. Will you come over? There's plenty of room in the stable and Susan's nosebag is highly recommended! Just say the word and we'll shake up the straw.

Regards,

Larry.

PS It is *important* that you come, not just because Susan laid me two to one you wouldn't . . .

CHAPTER FORTY

THE HEAVY RAIN BEATING ON THE CORRUGATED ROOF OF TALLYDUFF
Temperance Hall drowned the spatter of applause that greeted
the speaker. The Reverend William Barden had counted the audience
before he rose to speak. Twenty-eight. And himself. Twenty-nine.
Hardly an auspicious start, but a start. As he cleared his throat to
speak, a deep rumble of thunder rolled directly overhead. William
Barden took his cue. He raised his arms and looked to the leaden sky
before facing his audience.

"Friends, the Lord is with us!" A murmur of approval ran through
the hall.

"And if the Lord be with us, who shall stand against us? I thank
you for coming in such numbers to this inaugural meeting of the
United Church of Ulster. I thank you for answering the call. It is very
easy to ignore a call to the defence of one's principles, even of one's
faith. It is very easy to turn away and not, as St Paul said, 'Stand fast
and hold the traditions which ye have been taught.'

"The United Church of Ulster will stand fast – against the vipers
that insinuate themselves into the garden of Ulster, the serpents that
represent evil. We have witnessed over the past decade acts of
violence and barbarity carried out by an army from a foreign state
which purports to lay claim to the soil of Ulster – a foreign state
which is dominated and ruled by the Church of Rome. We who are
the seed and breed of Ulster are seen as the invader. We who live by
faith in God and allegiance to our Queen are under threat from the
real invaders who in their greed seek to dominate the entire island

302

and deny us our birthright." He paused to allow his words to sink home.

"Friends, when I am troubled I turn – as I'm sure you do too – to the word of God. And last night when I took down the Holy Bible to seek words of resolution for all of us, it fell open, propitiously I would say, at the Book of the Prophet Ezekiel, Chapter 16, in which the Lord calls on Ezekiel to let Jerusalem know of her abominations, her monstrous whoredom. Verse 26 says, 'Thou hast committed fornications with the Egyptians, thy neighbours and hast increased thy whoredoms, to provoke me to anger.' Verse 28 – 'Thou has played the whore also with the Assyrians, because thou wast insatiable; yea, thou hast played the harlot with them.'

"But the Lord will punish Jerusalem. Verses 40 and 41. 'They shall also bring up a company against thee, and they shall stone thee with stones, and thrust thee through with their swords. And they shall burn thine houses with fire, and execute judgements upon thee.' I say to you, friends, that the Lord will deliver similar judgements on the harlots of Dublin and Rome!"

He stood back from the table to acknowledge the prolonged staccato of applause . . .

"Will Daddy be home in time for the story?" Moira asked as her mother brushed out her long blonde hair.

"No. Daddy's away to a meeting," Hannah replied.

"What's the meeting about?"

"It's about the defence of Ulster."

"What's defence?" asked the boy who sailed a nailbrush across the basin of water.

"It's keeping Ulster safe, Roy, and will you kindly put on the top of your pyjamas."

"Is someone going to take Ulster away?" Roy asked.

"No, dear. Of course not."

"Daddy said the IRA were trying to take it away," Moira announced.

"Will they hurt Daddy?" Roy asked as he struggled into his pyjama top.

"Of course not," his mother said, turning her attention to the boy's hair. "Daddy's a good and just man. 'The just shall live by faith,' St

303

Paul said. Now into bed and I'll read you the story of the wise and foolish virgins."

"Any – way," Roy chanted as he bounced up and down on his bed, "Daddy's – got – a big – gun – und – er the – stairs."

"The just shall live by faith, my friends. In conclusion, let me remind you what we stand for. We are henceforth the United Church of Ulster – We are United – in our loyalty to God and Queen and in our defence of that loyalty. We are a Church – as Paul said to the Romans, who shall separate us from the love of Christ? – and we are of Ulster, by right and by entitlement!"

The small gathering spontaneously stood up and applauded the speaker warmly. William Barden acknowledged the applause and then held his hand up for silence.

"Thank you, friends. We have started on the road this evening. It will be a long and hard road but we have started. We may be few in number but remember – Our Saviour started on his road with less than half that number."

Brian Johnston stood, uncertain, amid the swirl of passengers alighting from the boat train in Euston Station. It was his first time in London and he was unaccustomed to the level of noise and bustle all around him. Suddenly, in the midst of all the clamour, a familiar Irish midlands burr came to him.

"De profundis Sussexii clamavi ad te – "

He turned to face Larry Rooney. His appearance had changed dramatically since their last meeting. His hair had turned white and his face had the wizened look of a man twice his years, but there was no mistaking the impish grin or the jaunty walk as he approached Brian. He paused for a moment, tears glistening in his eyes, before throwing his arms around Brian and hugging him fiercely. He kept muttering words Brian could only half hear.

"Good man, Brian – solid to the core – Rooney and O'Brien."

He released his grip and stood directly in front of Brian, smiling through the freely flowing tears. He took a sudden big breath that startled Brian.

"Jesus Christ! Jesus Christ! What will they think? Here's me hugging a priest openly and bawlin' me eyes out! Come on and we'll have a cup of coffee – unless you'd prefer something stronger?"

"Coffee's fine."

They found a quiet corner in a café.

"God, it's good to see you!" Larry said.

"You took the words out of my mouth," Brian replied. "You're looking great."

"'Deed and I'm not! Look at me! Old before my time. This marriage business would put years on you!"

They laughed together and sipped their coffees.

"Seriously though," Larry went on, "I'm sorry about my performance out there – "

"Nonsense."

"It's just when I saw you there in the clerical garb, the whole of Euston station dissolved and I could see you back in the seminary – and Stevie, laughing – and myself – " He choked back a sigh. "Christ, Brian – it only hits home at a time like this – "

"I'm sorry – "

"Don't be sorry. It's not your fault. It can happen with lots of things – Stevie's memorial card, a snatch of a song, a church bell – "

"But you're happy. I can see it in your face."

"Yeah I'm happy. Did you ever read Damon Runyon, Brian?"

"No."

"Ah you should. He has some great lines. My favourite is 'All life is six to five against.' I'd say he got it just about right!"

Brian smiled wistfully.

"Ah, will you look at me – blubbering and blabbering on like this!" He sat upright. "Now tell me all. How is life in Drumcash? And are you still entertaining 'doubts'?"

Brian told him of the lost weekend in Dublin and his move from Drumcash. Larry gave a low whistle.

"You've had a rough old time, friend. They didn't even give you odds of six to five. Listen – we'd better get out of here just in case they tailed you to London!"

Brian looked around nervously.

"Only jokin'! Only jokin'! We have a two-hour drive ahead of us. You can give me a blow by blow account on the way."

"Welcome to The Willows!" Susan extended a hand that was work-worn to Brian's touch. A compactly built woman, Susan's expression was initially dour and forbidding, but her smile transformed her into a warm and welcoming woman.

"I suppose Rooney has painted a picture of me as the dark evil one who stole him from his Church – "

"Absolutely! The Witch from the Willows!" Larry interjected. "She'll be burned at the stake if she ever goes back to Balgower!"

"Have you ever gone back?" Brian asked.

"Not a chance. I'm bad news still. Just as well Meath haven't made it to the All-Ireland since!"

The Willows was an old Tudor-style farmhouse with low ceilings and squeaking floors. It had a warm, lived-in feeling about it and Brian felt instantly at ease.

"Well, are we going to feed this poor man at all? He didn't come all the way from Drumcash to admire the view!"

Brian looked across at the rolling downs of Sussex.

"It is quite a view."

"Right. To work!" Susan clapped her hands. "Hope you like pork, Brian."

"We killed the fatted pig," Larry whispered.

"Rooney, pour Brian a sherry. Usual for me. You remembered the bitter lemon, I hope."

"Jesus! I clean – "

"I should have known when you two seminarians got together my basic needs would come a poor second. Never mind. I'll put up with the tonic – for now."

"Very understanding woman, Susan," Larry said. "I think that's what attracted me to her in the first place!"

"What attracted you to me, Rooney," Susan called from the kitchen, "were my fillies, my fortune and my figure – in that order."

"Has a cruel streak in her, but I'm the forgiving type!" Larry's chuckle was echoed by the sherry bubbling from the bottle.

Over a long and relaxed meal, Susan and Larry entertained Brian with an often hilarious account of their furtive courtship.

"Courting in a horsebox – with the horse present – is not recommended," Larry sighed.

"Yes, I think when the horse farted and Rooney blamed me – I think when I forgave him that, I was doomed," Susan added in a resigned voice.

"I was doomed when my photograph appeared in the *Irish Independent*," Larry said. "I'm told poor Father Fay nearly choked on his boiled egg!"

"Why so?" Brian enquired.

"Because I was photographed in the Members' Enclosure at Aintree standing behind Vincent O'Brien – "

"And beside me," Susan reminded him.

" – when I was supposed to be on a three-day retreat at Knock! That's why!"

Towards the end of the meal, Susan began glancing anxiously at her watch.

"Come on, Rooney," she said suddenly after a lull in the conversation. "Haven't you told Brian the real reason you asked him over? I'm afraid it wasn't just to meet The Witch from the Willows, Brian!"

Brian was genuinely puzzled. Larry addressed him in a hushed, serious tone.

"You see, we're organising a major betting coup at Chester and we were looking for a respectable front man – "

"Ah, Rooney!"

"Sorry, sorry. Joke! Joke! Horsey Protestants have no sense of humour, Brian." Larry was almost bowled over by a well-directed blow from a cushion. "Seriously – a couple of months ago I discovered an Irish priest who ministers about ten miles from here. I got talking to him and – lo and behold – doesn't it turn out that he taught you at St Felim's!"

"Taught me? Who? – "

"His name is Vincent Duffy. He said you'd remember the soirées – he hoped!"

"Vincent? Father Vincent? Here?" Brian was incredulous.

"Not only that but he's due here any minute," Susan announced. "We did ask him to dinner but he had a prior engagement – at a concert in London. Shall we withdraw?"

She motioned towards the adjoining room.

"We have our soirées too, Brian. Father Vincent comes over occasionally to play for us." She gestured towards an elegant piano whose polished black surface caught the reflection of the setting sun.

"A wedding present from father. Family heirloom."

"It's a beautiful piece," Brian said.

"Rooney was disappointed. He was expecting a brood mare."

"Dead right," Larry said. "Sure there's no breeding potential there." He faced up to the piano as if he were in a bloodstock sales ring. "Bad conformation and terrible legs." He winked at Brian.

"No soul, Rooney. Absolutely no soul," Susan said disparagingly as she lit a cigarette.

"Ah, but lots of heart, dear. Lots of heart!"

Brian turned away in embarrassment as Larry squeezed his wife's buttocks.

CHAPTER FORTY-ONE

THE LONG SLENDER FINGERS CARESSED THE PIANO KEYS, COAXING A FRAGILE Mozart melody from them. Brian closed his eyes. He might well be in Father Vincent's room in St Felim's. Father Vincent paused momentarily and threw a severe look at Art McGrane who was chomping his way noisily through the Marietta biscuits. He could hear Conor Maguire snorting behind his hands at Art's misfortune. Father Vincent resumed playing . . .

"Now there's a contented man," Larry said loudly.

"Shush, Rooney," Susan snapped.

Brian's reverie was broken. He looked across at the pianist. Father Vincent had changed very little in almost twenty years. The hairline had receded but otherwise he was the same Vincent – a slight figure with a tidy appearance. And those long slender fingers.

The music ended. The audience of three clapped in appreciation.

"Penny for your thoughts, Brian," Susan said.

"You couldn't buy them." He smiled. "Just cherished memories of another time."

Vincent sat beside him on the settee.

"I often thought you were the one who appreciated the soirées most," he said. "The others were only there for the biscuits."

"No soul – just like our genial host." Susan jabbed the dozing Larry in the arm.

"The cloth suits you, Brian," Vincent said. "You'll have to come over to Catherton so I can show off my past pupil."

"Keep him away from eligible horsy women," Larry jibed. "They ruined many's the good man!"

"You'll need somewhere to say Mass anyway," Vincent continued.

"Rooney will drop you over," Susan said.

"Aye. No problem. I'll even serve for you – just to keep my hand in!"

"I promised you it would be worth the climb! We can rest here." Vincent seated himself on the grass at the top of Cather Tor. Brian stood for a few minutes admiring the undulating parkland that lay before him. Pockets of mist hung lazily over the lower parts of the valley before the heat of the day would dispel them. In the distance a string of racehorses cantered silently across a huge railed field before disappearing into a dip.

"Makes a change from the wee hills," Brian said as he seated himself.

"They have their own beauty," Vincent said, propping himself on one elbow. "I miss them."

"Do you ever go back?"

"No. No reason. No ties." He turned to face Brian directly. "Are you happy, Brian?"

Brian shrugged his shoulders. "What is happiness?"

"Indeed – and what it truth? What are you smiling at?"

"I just remembered Conor Maguire telling you that you didn't look happy one evening."

"And – ?"

"And you withered him with a look saying 'Haven't I the right to be unhappy in my own room?'"

Vincent laughed. "Did I really? Poor Conor. Whatever happened to him?"

"The last I heard he was anything but poor. Made a fortune on the buildings over here after the war."

Vincent plucked a long stem of grass.

"At the risk of getting a withering reply, may I suggest that there's a lot of pain in your life, Brian. It's written in your face."

Brian looked away. The string of horses had re-emerged from the dip.

"Is the writing that clear?"

"Yes."

Brian gave a deep sigh. "Yes, of course there's pain. Loss. Separation. It's in everybody's life to some degree."

"Tell me about yours."

His concern was so evident and his tone so reassuring that Brian found it easy to unburden himself of much of his pain – how Hannah and he had drifted apart. Was it an aimless drifting or had both their circumstances driven them apart? Now there was the latest news he had heard about William Barden. Then there was Sheila and his move to St Felim's Hospital.

"There's one other source of pain, but it's something that must remain with me and which I must deal with myself."

Vincent remained silent for a long time.

"Although I didn't think so at the time, my being sent to England was a godsend. I worked under a big blustering, forthright and very wise Kerryman. 'Don't bother about seeking for happiness,' he used say. 'Happiness is shallow, ephemeral. Joy is what we should be after. And it is within our grasp. Joy is about transcending the pain of living, whether that pain is physical, mental or spiritual. That's why the Christian story is a metaphor for life. We come through difficulty to enlightenment. We die to ourselves in order to recreate life.' Am I making sense?"

Brian nodded.

"Here endeth the lesson for today," Vincent said, springing to his feet. "The one thing you're guaranteed up here is a good appetite!" He glanced at his watch. "And Mrs Buxton's lunch beckons."

The week passed all too quickly. Brian grew quite fond of Catherton – a pretty village whose Catholic congregation Vincent described aptly as "very polite and very civilised." Susan and Larry smothered him with hospitality. The more his acquaintance with Susan deepened, the fonder Brian grew of her. Behind the brusqueness of manner and speech, Brian discovered a woman of great warmth and affection and a wife who loved her husband deeply.

Larry never ceased to amaze Brian. They took Brian to Sandown Park races on the Saturday. He marvelled at how popular Larry was among the racing "set". He remained his natural self, his ready wit and laconic comments endearing him to all in his company. On the way home, Susan and Larry shared their secret with Brian. Susan was pregnant. .

"I'm delighted for you," Brian said.

"At least I can style myself Father Larry again!" Larry chuckled.

Parting was difficult. He said goodbye to Vincent after Sunday Mass.

"I hate goodbyes," Vincent said.

"I remember – from the last time," Brian replied, recalling Vincent's sudden departure from St Felim's.

"Just remember. Seek for joy – and keep in touch. It was good meeting you, Brian." He offered his hand in farewell. Brian took it in both of his.

"Thanks," he whispered. Susan and Larry drove him to Euston Station.

"You take care of yourself," Susan said as she kissed him on the cheek. "Good priests are hard to find."

"Most of them got married," Larry quipped.

"Did I or did I not do the people of Balgower a favour by snatching this Philistine from their midst?" Susan sighed.

Larry hugged him. "Goodbye, old stock. Remember – as long as they keep offering you six to five, take it!"

As the train pulled out of Euston, Brian reflected that, each in his own way, both Vincent and Larry had given him the same advice.

PART THREE

CHAPTER FORTY-TWO

APRIL 1975. BRIAN JOHNSTON RETURNED TO THE PRESBYTERY AFTER THE Sunday midday Mass. He was tired. It had been midnight before he had returned home from visiting Hannah – and he had not slept well.

He made himself a cup of tea and was just about to enjoy it when the telephone rang.

"Uncle Brian?" He was taken aback at hearing Moira's voice so soon.

"Moira! Is she – ?"

"No. No. But she has gone down very quickly since you left. The doctor thinks it'll not be long now. I thought I should let you know."

"Thanks, Moira."

"There was one other thing. Last night she was rambling a lot. Kept calling me Aunt Rose, saying, 'Look, Aunt Rose. Look what Brian won at the sports. Can I go to Drumgore every year?' In fact they were her last words. She's been too weak to speak since then. I just thought you might – "

"Thank you, Moira. You're very good to ring. I appreciate it."

"Well, I'll let you know if – "

"I know. God bless you, Moira."

He sat there in the hall for a few minutes. When he remembered the tea it had gone cold. He sighed and switched on the radio to catch the lunch-time news. He would make a fresh cup of tea. The news bulletin opened dramatically.

"Six people are now dead after a spate of killings in Belfast over the weekend. Several people have been injured, some critically . . . "

Brian hammered the draining-board of the sink in utter frustration. How long would it go on? For five years now the spiral of violence had been unrelenting. The reporter's voice drifted back. " . . . The dead have now been named . . . " The dead . . .

An intensely humid July afternoon, 1966. His fortieth birthday had just passed. A small group stood shrouded in fine drizzle in the *cillín*, the little field across the road from Cullyboe cemetery. Hughie and Josie McKevitt, their sons Peter Joseph and Charlie, Bridie and Brian. Hughie and Peter Joseph had opened the McKevitt grave in the cemetery while Brian and Charlie opened Sarah's grave in the *cillín*. Bridie had stayed in the church while the digging proceeded. Brian and Charlie worked steadily in the oppressive heat until they came upon fragments of wood.

"We'll go easy now," Brian said. They picked gently at the yellow clay until the first bone appeared. Brian cast the shovel aside and scooped and brushed the earth aside with his hands. The skeleton was almost intact. With trembling hands he brushed the remaining earth away. He ran his fingers lightly along one bone. It was at once the most unreal and the most intensely private moment of his life. The drizzle washed the remaining grit from the bone, exposing its fine graining.

Running his fingers underneath the bone, Brian dislodged a fragment of cloth. He slipped it into the pocket of his overall. Josie had prepared a pallet strewn with flowers to receive the bones. Charlie and Peter Joseph carried the pallet across the road to the cemetery, while Brian read aloud the prayers for the dead. Before they lowered the pallet into its new resting-place, the little group huddled around the open grave and prayed in silence. Josie draped a patchwork quilt over the remains. As the men filled in the grave, Bridie led the recital of the rosary.

"She's at peace now," Hughie said, wiping the sweat from his brow with the back of his hand.

"She was always at peace," Bridie said, still fingering her beads.

"Still – "

Brian withdrew the fragment of material from his pocket and handed it to Bridie. She stroked it reverently as it lay in the palm of her hand.

316

"She looked so lovely in that dress. Canon McKenna didn't want her to be buried in it but I – " Her voice began to choke with emotion. Josie put her arm around Bridie, who kept staring through tear-filled eyes at the faded and smudged piece of cloth.

Man in the moon, man in the moon,

Why did the dish run away with the spoon?

Brian went back to Bridie's house for tea.

"I can't stay too long," Brian said. "I have an appointment at eight o'clock."

"It's just a bit of cold meat. I'll not be long." She rooted in a drawer for cutlery. "Have you settled in to Kilcartin?"

"It's grand. A nice wee place. Have you ever been there?"

"No. Podgy had a cousin lived over that way. He's dead this few years. Never cared much for him."

Brian smiled as he scanned a newspaper.

"'There's nothing smart in Kilcartin,' Jimmy McCabe used to say," Bridie continued.

"Do you ever hear from Jimmy, now?"

Bridie busied herself cutting soda bread. "Och, indeed. He's a terror with the pen. Always asking me to go out to the States."

"Would you not go?"

"Not at all. At my stage in life?" She placed a salad in front of Brian. "Anyway from what I see on the television I wouldn't want to visit yon place. It's a wicked country, full of crime. And as for the women. Not an ounce of shame in them. Och, I forgot the scallions." She was up and out the door to the vegetable garden before Brian could tell her that he didn't like scallions . . .

He reflected on Jimmy McCabe's comment – "There's nothing smart in Kilcartin" – with a smile. He had been appointed curate there a year previously. It was good to be back in a parish again. Seven years in St Felim's had been enough. His parish priest, Jim Reilly, an ailing man in his seventies, was content to serve out his days untroubled by the deliberations of the recent Vatican Council. It fell to Brian to begin to implement the liturgical changes in Kilcartin – a slow and difficult process. The people of Kilcartin showed little routine interest in the laity's participation in the liturgy. Brian had decided to work through the children who were less inhibited than the adults and he was enjoying taking them through the new liturgy of the Mass.

Bridie dropped two enormous scallions on his plate. "Now! There's more if you want them." She drew a chair up to the table and began to eat. "I declare to God I was in the church above when Matty Martin and the son came in and started measuring out for a new altar. They tell me it's just going to be a plain table. Sure what's the need of that and a beautiful altar there behind them, that's there this hundred years?"

"It's part of the new approach. Involving the people more in the Mass. The priest will be facing you from now on."

"There's many of them would be as well staying as they are!"

Brian was not going to be drawn. He stole a glance at his aunt. Her furrowed brow and occasionally shaking head indicated her displeasure with the pace of change that was clearly too fast for her liking.

"Have you enough scallions?" she asked.

"I'm grand, thanks." He was still struggling with one of the two on his plate.

"It was my father's pennies – and his father's – that built that altar. And now it's not good enough any more. It's a quare world."

The Angelus bell from Cullyboe rang out in the misty air. They recited the Angelus together before turning on the television for the news.

There were reports on several twelfth of July parades. The camera homed in on a familiar slightly-built figure dressed in clerical grey and wearing a bowler hat.

"Well, will you look at the cut of thon!" Bridie exclaimed. The Reverend William Barden, Chairman of the United Church of Ulster, was being interviewed.

"It is very important for us to show our strength and solidarity with the Union in the light of all the triumphalism and sabre-rattling we have witnessed in the south this year."

"You are referring to the Golden Jubilee celebration of the 1916 Rising?"

"I am referring to the parading and posturing that is little more than a veiled threat to the stability of this province – "

"There are those who would say that the twelfth of July commemoration is equally a matter of parading and posturing – "

"That, sir, is but a betrayal of your allegiance to another state. We are simply stating that we are of Ulster, we are of the Union, we are loyal to the Crown and we will have no truck with Dublin or Rome."

"That sort of talk will do no one any good," Bridie said. "I don't know how Hannah got involved with his like."

"Nor do I," Brian sighed. "Nor do I."

He arrived back in Kilcartin shortly before eight o'clock. A fresh-faced young girl sat alone in the sitting-room of the presbytery, fidgeting with her fingers. Maureen Waters was an orphan. She had been reared by her Aunt Anna, a deeply religious woman who always wore black, thus earning the sobriquet "The Widow Waters" – despite the fact that she had never married. Maureen was a shy, introverted girl who spoke in a subdued voice and always averted her gaze when spoken to.

She withdrew an envelope from her cardigan pocket. "Aunt Anna said to give you this," she whispered as she proffered it to Brian. He tore the envelope open.

Dear Father Johnston,

I cannot bring myself to discuss this child's state. Suffice to say I am agreeable to an arrangement being made with the Rogans.

Yours sincerely,
Anna Waters.

Brian glanced across at "the child" sitting nervously on the edge of the chair. Her "state" was that she was pregnant. Joe Rogan had forced himself upon her when she had brought him tea during haymaking. In her innocence, Maureen had told her aunt all about the incident. Brian had been asked to find a solution to the problem.

When the Rogans arrived – father, mother and son – the solution was offered instantly.

Charlie Rogan was a squat, compactly built man, whose body seemed as if it had been sown tightly into his skin. In his footballing days he had been known as "Rubber" Rogan. Even now, if he were to fall over, it seemed he would surely bounce back up again.

"I won't beat around the bush, Father," he began brusquely. "We'll do the decent thing and make a decent woman of her." He never looked at Maureen. His son stood sheepishly behind him, hands sunk deep in his pockets. His wife sat, uncomfortable and sullen, at the other end of the room from Maureen.

"He'll marry her, the sooner the better. Might put a stop to his gallivantin'."

"She's a decent woman already, Charlie."

"That's as may be, but she's in trouble, isn't she? The Rogans always paid their dues – to church and chandler."

"In the name of God, Charlie, this is not the fair of Carrick. You're not bidding for the girl."

Charlie Rogan gave an impatient toss of his head and stepped forward to face Brian directly.

"Have you anything else in mind, Father? Because if you're not interested in what I'm saying – " His voice trailed away in a whiskey breath.

"All I'm saying is that marriage is a big step. We don't know what the young people themselves think – "

"Well, ask them!" Charlie swung his arm in a sweeping gesture towards them.

Joe Rogan shrugged his shoulders and smiled. "I'm game," he said.

Brian turned to Maureen. "You could have the baby adopted – "

She turned to him suddenly with a look of terror in her face.

"No. I want to keep my baby. I'm getting married," she cried in a hoarse voice.

"There y'are!" Charlie crowed. "They'll be grand, I'm tellin' ye!"

Brian looked from Charlie Rogan's face, its skin so tightly drawn that he felt it must burst at any moment, to the letter which lay on the table before him. "I am agreeable to an arrangement . . . " He felt defeated, as if he had personally sold the trembling girl to the only bidder . . .

A week later at eight o'clock in the morning Joe Rogan and Maureen Waters were married. In that same week Brian had received news of another impending wedding.

Kilcore
Cloyne
19th July 1966

Dear Brian,

I hope you are keeping well and are enjoying your new parish. How are you coping with Vatican Two? Our PP absolutely refused to give Holy Communion into the hand – so much for progress down here!

Mother is confined to bed all the time now. She seems to have grown smaller and smaller and looks like a little sparrow in the bed.

The big news is that I have become engaged to Noel O'Sullivan. You may remember I mentioned him some time ago. He has a farm about five miles from here. He's a nice lad. You'd like him. We have no wedding date in mind yet, because of mother, but I'll keep you informed. Take care of yourself.

Sheila.

PS Any news of Larry?

Brian folded the letter neatly. Another chapter in his life had closed.

He had no news of Larry then but a year later Meath's participation in the All-Ireland Football Final inevitably brought Larry Rooney home. He presented himself on Brian's doorstep one afternoon in late September.

"I can't say I'm surprised to see you." Brian laughed. "I was half expecting you. Congratulations on another win!"

"Thank you." Larry's voice was little more than a whisper. "The voice is a bit delicate! Yes, it was a long wait – thirteen years – but all the sweeter for that! And it was victory on all fronts. Would you believe I showed my face in Balgower and I wasn't shot. I was actually made welcome!"

"I'm glad," Brian said. "What are we standing out here for? Come in! Come in! I'll put the kettle on – or would you like – ?"

"A cup of tea would be grand."

"Is Susan with you?"

"No. The children would be too much of a handful – "

"Is it two or three?"

"Two. Boy and girl. Typical Protestant family!"

Brian fussed around the kitchen searching in presses and drawers.

"Don't bother with the good china!" Larry jibed. "A mug will do me fine!"

"I was just looking for a biscuit. There's a packet of digestives somewhere."

"Don't bother. Sit down there and tell me all. Changed times. No housekeeper, what?"

"She comes in to cook the dinner and tea."

"I don't know if I'd come back on these terms!" He winked as Brian looked up in surprise. "Don't worry. I'm too fond of the home comforts at this stage."

"I noticed you weren't undernourished!"

Larry patted a bulging waistline. "Carrying top weight these days. The handicapper hasn't been kind to me." He strolled around the kitchen peering out through the lace-curtained window. "Big changes all round, Brian. How are you coping?"

"The people of Kilcartin will never rush into anything new. The older people are not too keen on the Mass in English. I concentrate on the younger generation. And there's an almighty row over taking away the altar rails!"

"Listen, old friend. Don't worry about altar rails! The big problem you're going to have is not English Masses or handshakes. It's the pill."

There was an uneasy silence. Brian drained his mug with a large swallow of tea. "We're waiting for the encyclical," he said timidly.

"Ach, you know what that will say. 'Thou shalt not contracept!' I'm telling you. Today's young people won't buy that."

"Are you speaking from experience?" Brian laughed nervously.

"Of course. How else do you think we managed a Protestant family?"

Brian rinsed out the mugs. "How is Vincent?"

"Not great! Not great! Crippled with arthritis. Can't play the piano any more. That's what really kills him. I suggested selling the piano and buying a yearling. I can tell you the pill didn't matter that night!"

"Poor Vincent," Brian said. Larry looked out another window.

"God in heaven. No matter where you look there's nothing but hills around here. Does it not drive you mad?"

CHAPTER FORTY-THREE

ALTHOUGH IT WAS ONLY MID-AFTERNOON ON A SULTRY JULY DAY, WINSTON Pollock drew the heavy curtains in the drawing-room of Sloe Hill and joined the three men seated at the table. He introduced the man on his right.

"Gentlemen, this is Walter Hamilton. The Reverend William Barden, Mr Samuel Jackson." Handshakes were exchanged. "Mr Hamilton is in the shipping business. Next month he will be landing a cargo that will be of interest, and more to the point, of use to the Sons of Ulster. I don't think any of us need reminding that these protests about housing and civil rights are posing a threat to the stability of this province. We all recognise them for what they are – "

"A front for the Republican Movement," William Barden interjected.

"Precisely. That is why we need to be on our guard and to be prepared to defend our heritage. Mr Hamilton here will now give us details of his recent acquisitions."

Before Walter Hamilton could speak, Hannah Barden entered the room with a tea-tray. Winston Pollock glowered at her husband as Hannah distributed cups and saucers. Walter Hamilton began to speak but Winston stayed him by clamping a massive hand on his shoulder.

"Thank you, dear," William Barden said. Hannah smiled in return and left the room without speaking.

"Was that wise?" Winston muttered.

"What was unwise about it? Hannah and I have no secrets from each other. She is on our side." William Barden's reply had an edge of resentment.

"She has blood ties," Winston snapped. "Now, Mr Hamilton. Continue."

"Chuck-chuck! Chuck!" Bridie McKevitt winced as she bent to look under the briar-tangled hedge in search of the straying hen. She poked with her stick wherever a path appeared among the nettles. "Chuck-chuck! Come on, my wee lassie. I know you're laying out somewhere hereabout. We've been watching you for a week now – me and St Anthony – and you're not doing my old back any good with all this stooping. Chuck-chuck!" Every step she took sent a searing pain through her right hip. Her patience was fast diminishing as the pain increased. She edged nearer the end of the hedge. "St Anthony, St Anthony, find me my luck," she whispered.

She cut a swathe through the nettles to reveal the Rhode Island Red squatting on a crude nest under the innermost part of the hedge. "Aren't you the bold lassie causing me all this trouble!" Bridie poked the stick through the puffed-up feathers. The hen moved away reluctantly to reveal five eggs in the nest. "God bless St Anthony! Never failed me yet!" Bridie panted as she sank painfully to her knees. She could feel the thorns piercing her cardigan sleeve as she stretched her arm in the direction of the eggs. Her face brushed against the tall grass. Suddenly a heavy black shutter shot violently across her eyes with a ringing metallic echo. All was black and intense, intense pain . . .

The black dissolved into a radiant white core with a soft green beyond. They were seated in the summerhouse in the orchard – James, cigarette askew between his lips, idly scanning a newspaper; Pete, leaning his head back into cupped hands; Hughie, looking away into the distance as he chewed a stem of grass and Bridie, darning a sock, holding it up to the light every few minutes to check her handiwork.

"Begod!" James exclaimed suddenly. "Podgy'll not like this!" He read from the newspaper. "'Peter Kernan, publican, was given the benefit of the Probation Act when he was charged with allowing a number of men to be present in his premises in Drominarney after midnight on May 16th last. Defendant caused laughter in court when

he said the men had come in after the closing of the mission, having been praying all week. They obviously didn't take the pledge at the mission, Justice Carter said, amid further laughter.'"

"He'd talk his way out of hell, that fellow," Pete said.

"I was down there the other evening," Hughie said, "and he says we must have notions about ourselves putting up a summerhouse in the garden!"

"The ould eejit – " Pete began.

"And he wanted to know – " Hughie winked at his sister " – what did Pete have to paint it white for. It has him blinded when the sun is on it, he said."

"Too bad about him! I'll change it to duck-egg blue in the morning," Pete muttered. "He could do with a drop of paint on that place of his. Stingy ould get!"

Hughie suppressed a chuckle.

"Och, he's not the worst in the world," Bridie said quietly.

"Be the holy!" Pete replied in genuine surprise. "Jimmy McCabe is going to get his eye wiped if he's not careful."

"There's no eye to wipe," Bridie said with more than a hint of annoyance.

There was a tense silence until James broke in again.

"Here it is!" 'Cullyboe fall at the first hurdle,'" he read. "'Cullyboe crashed out of this year's Feis Cup when they were surprisingly beaten by Kilcartin at Castleblayney. The cupholders never recovered from a shaky start when they conceded two soft goals in the first ten minutes. Despite some majestic fielding by McKevitt at centre-field and some incisive runs by McCabe at left-half forward, Cullyboe were a major disappointment and at the end were a well-beaten team.'"

"A major disappointment is right," Pete said bitterly. "You might sing that! Two soft goals!" He spat the words out. "I'll tell you. Marriage didn't do Spud Tallon's football any good. Dropped two high balls that me granny would catch!"

"His mind wasn't on the job!" Hughie laughed.

"Brian obviously gave the Kilcartin boys a special blessing," Bridie said.

"He tells me he's going to be on television in a couple of months' time," James remarked as he folded the paper. "They're going to televise Mass from Kilcartin."

*"God, there'll be no living with them, between that and the football,"
Pete sighed. "Curse of God on them anyway. It's going to be a long
summer around here." He stood up and, shielding his eyes from the
sun, he gazed with growing curiosity in the direction of Flaxpool Lane.*

*"There's someone in an almighty hurry. Begod, it's Sarah!" The
other three joined him in watching their sister's furious progress as she
cycled towards them, head bowed, her floralprint dress flowing behind
her. They moved as one to the gate to greet her. Sarah dismounted and
let her bike fall on the roadway.*

*"What is it, Sarah? What has happened?" Bridie moved towards her
distressed sister.*

*"It's William Barden," Sarah sobbed. "He has taken wee Hannah
away. He said he was going to look after her from now on. Said I
wasn't a fit person to mind her. Said the Lord had told him to take
her . . ."*

*"Begod we'll see about that," Pete muttered as he clenched the top
bar of the gate.*

"Did you tell Canon McKenna?" James asked.

*Sarah dabbed her reddened eyes with the frill on the sleeve of her
dress. "I did, but he just gave out to me. Told me it was my own fault.
Said he had warned me. Then he just turned his back on me."*

*Her slight figure trembled visibly. Bridie put her arm around her sister's
shoulders. "Come on inside and I'll make you a nice hot cup of tea."*

Something was tickling her hair. She turned her face sideways and
found herself eye to eye with the Rhode Island Red.

"Bad cess to you," she muttered. The hen tilted her head
downwards as if expressing disapproval of Bridie's comment. Bridie
turned her head in the opposite direction with great difficulty. She
raised her right hand. A sticky clinging mess of egg-white and egg-
yolk dripped from her hand onto the crushed shells. Four of the eggs
were broken.

"Bad cess to you again!" she said to the hen which had now
wandered away. "Them eggs were for Sarah." She wiped her hand on
the long grass, picked up the remaining egg and began the slow and
extremely painful process of hauling herself first to her knees and then
to a standing position. Her whole body ached as she leaned heavily
on her stick and took the first tentative step back towards the house.

"Sarah'll make me a nice cup of tea. I'll be grand then. Grand."
Progress was slow. She began to hum a tune repetitively.
"Man in the moon, man in the moon,
Why did the dish run away with the spoon?"

"I hope Jack Lynch is watching this," Father Jim Reilly said as he and
his curate Brian Johnston watched the television reports of rioting in
Belfast. It was August 15th 1969. The television news carried
disturbing pictures of death, injury and destruction on the streets of
Belfast. "He can't stand by any more like he said the other day. He
has to send the Irish Army in to help those poor people." The longer
the report went on, the more agitated he became. "They're reaping
the whirlwind now! Bernadette Devlin was right. It's not one night of
broken glass but fifty years of human misery."
    "Maybe the British army will calm things down," Brian said.
    "Not at all. They're the British army, aren't they? – the army of the
oppressor. They're not going to have any great sympathy for our
people. It's time for all Catholics to stand together."
    Brian feared for his parish priest's health. The high colour had
returned to his face. His hands trembled as he attempted to sip from
a cup of tea. Apart from his concern for the older man's health, Brian
was surprised by his reaction to the trouble in Belfast and earlier in
Derry. The man had shown no enthusiasm for the changes that had
come about in the Catholic Church after the Vatican Council. He had
practically ignored the liturgical changes, leaving it to Brian to
implement them. The violence in Northern Ireland was another
matter. It had awakened deep emotions in him, emotions that were
not going to be easily assuaged.
    The telephone rang in the hall. It was a welcome distraction for
Brian. He didn't recognise the voice at the other end at first. It was a
very upset Josie McKevitt.
    "Brian? Thank God I got you. It's Aunt Bridie. Charlie called in to
see her on his way to Dundalk. It was the grace of God that he did.
He found her in a heap on the kitchen floor. They've rushed her to
the County Hospital. They think she had a stroke." Brian was already
reaching for his coat on the hallstand.
    "I'm on my way," he said.

He edged his way between the screens around the bed in the corner of the ward. A beatific smile spread across Bridie's pallid features.

"Ach, it's wee Brian!" she whispered. "I have just the job for you. It's that wee rogue of a hen. She's laying out on me again and I can't find the nest. Brian's the boy'll find it, aren't you?"

"Of course, Aunt Bridie," he said, reaching into his pocket for the holy oils with which he would administer the last rites.

"Poor wee Sarah!" Bridie turned to the nurse at the other side of her bed. "She got a terrible bang on the head doing the roly-poly. Will she be all right, doctor?"

"She'll be grand, Bridie," the nurse reassured her.

"Look, Sarah! It's the man in the moon. Pull back the curtain till we see him rightly!"

Josie McKevitt watched as Brian gently dabbed Bridie's eyes with cotton wool.

"Through this holy unction and through his most tender mercy, may the Lord pardon thee . . . "

Josie turned away and buried her head in her hands.

Brian sat by the bedside for some hours. He turned occasionally from reading his breviary to gaze at the now sleeping figure of his aunt. Not his aunt, his mother. The mother who had reared him, protected him, loved him, shaped him. He looked at the pale hands now resting on the blue counterpane. Hands kneading dough, hands fingering rosary beads, hands cradling Pete's head among the potatoes where he had fallen . . .

"You really should go home and get some sleep, Father," the night sister whispered. "She could go on like this for a long time."

There was little change in Bridie's condition a week later when Josie brought a visitor to see her.

"Now, Bridie," she announced. "Look who has come to see you – all the way from America. It's Jimmy! Jimmy McCabe."

The dapperly dressed figure moved forward to take Bridie's hand.

"Hello, Bridie. How is every bit of you?"

Bridie shook her head slowly.

"Jimmy McCabe! You're a danger to man and beast the way you ride that bike."

Jimmy bit his lip.

"I'm home now, Bridie. Don't need the bike any more. I'm home, Bridie. Forty years late, but I'm home."

It was when Josie had gone to Bridie's house to tidy up that she had come across the letters all pressed neatly at intervals between the leaves of *The Works of Robert Burns*. The personalised notepaper bore Jimmy's telephone number and address. He had caught a plane on the same day that Josie had rung.

For the last seven days of Bridie's life, he spent most of his waking hours at her bedside. He was intrigued by *The Works of Robert Burns*. Each time he opened it, the book fell open at page 120 – "Thou hast left me ever, Jamie."

"God, Bridie," Jimmy sighed. "Why are we so goddam foolish and so goddam proud?"

He whiled some of the time away each day reading poems to Bridie. "What about this one, Bridie?

When I think on the happy days
I spent with you, my dearie
And now what lands between us lie
How can I be but eerie?
And now what lands between us lie
How can I be but eerie?"

On September 1st 1969, Bridie McKevitt died.

A small group of family and neighbours assembled in the mortuary for the removal of her remains to Cullyboe. Hughie stood by the coffin, the last of the McKevitts, accepting the sympathy of those present – awkward, ill at ease, unsure of what to do or say. Jimmy McCabe gazed in disbelief at Bridie's china-like features, unlined and fragile. She had been transformed into the Bridie he knew forty years previously, save for the fine back-swept silver hair which he now stroked reverently. He closed his eyes and took himself back to a warm summer night when the scent of hay hung trapped beneath the blanket of mist that lay across Drominarney. He could hear the laughter and cheers of the revellers departing from the schoolhouse. Most of all, he could sense the warm, vibrant body of a young woman close to him and he could feel the electric tingle of her fine auburn hair flowing back into his face . . . He kissed the marble brow softly and withdrew.

Bridie McKevitt was laid to rest in Cullyboe cemetery. When the final prayers were said, Charlie and Micky McMahon spat on their hands, took a shovel each and began filling in the grave. The mourners slipped away, some pausing occasionally to pray at a family grave. Brian made his way to another McKevitt grave. The recently added black lettering at the bottom of the headstone glistened in the bright September sun.

And their daughter Sarah,

Died 21 December 1926

He knelt in prayer for a few moments.

As he left the cemetery, a tall figure dressed in an unseasonal long black coat approached him. He nodded and made to move on, hoping to avoid the embarrassment of not recognising someone from Cullyboe whom he had not seen since childhood.

"Hello, Brian." The voice was low but distinctive. "I just wanted to express my sympathy over Bridie." It was Rose Johnston.

Brian was taken aback at her care-worn appearance. Rose would be of an age with Bridie, but she now looked ten years older than her.

"That's – very kind of you, Rose. You were very good to come down."

"She was a lady. I always respected her."

"She was a lady, Rose."

"And she didn't have the easiest of lives. I know that."

"How is Hannah?"

"She's grand." Rose reached for the bicycle propped against the cemetery wall. "I only see her a few times a year. She has two lovely children, thanks be to God. Both away in England. Moira's nursing and Roy is studying law."

"Give her my love. I hope we can meet soon."

"I hope so too. It's not right that you should – it's not easy," she whispered.

"I understand," Brian said. "Can I give you a lift? There's a bit of food over at Uncle Hughie's – "

"Not at all. I'll be off home. Winston will be expecting his dinner."

"I could give you a lift as far as – "

"I'll be grand," she said, mounting the bike unsteadily. "Anyway, it wouldn't do to be seen by some driving around with a priest!" There was only a hint of humour in her remark.

"Ach, Rose – "

"Goodbye, Brian," she called. The long black coat flapped about her as she cycled slowly out of Cullyboe.

CHAPTER FORTY-FOUR

T HE HEALTH OF FATHER REILLY, PARISH PRIEST OF KILCARTIN, CONTINUED TO deteriorate and by 1971 he was confined to bed for most of each day. The worsening situation across the border did not help matters as each new account of rioting, British army searches, the closing of border roads and violent deaths upset him greatly. The entire burden of parish work fell to Brian. In the course of that work, he sensed a growing anxiety among his parishioners about the way things were developing only a few miles from where they lived. For some, that anxiety was transformed into calls for action. Slogans appeared overnight, painted on bridge walls or affixed to telegraph poles:

Provisional IRA for a New Ireland

End British Rule – Join the Provos

On a sultry May evening in 1971, the parish of Kilcartin experienced the reality of violence. An urgent knock at the presbytery door summoned Brian from his perusal of the scripture readings for the following Sunday. A distraught youth stood on the doorstep.

"You have to come quick, Father. There's been a terrible accident." He ran back to a red van which had its engine running.

"Will ye come with me or – ?"

"I'll follow!" Brian called.

He trailed the red van for some miles into the countryside, along narrow pot-holed roads that wound through the hills of north Monaghan. The van slowed and turned into the yard of a deserted farmhouse but then kept going through gateways and across rough fields. Brian was bemused. He had no option but to follow. Finally the van stopped outside a crude cattle shed in the corner of a field. There was no roof on the shed. Brian had to negotiate his way around sheets of corrugated iron that were strewn on the grass in the vicinity of the shed.

"In here, Father." He followed the youth into the shed. Inside was total carnage. Debris littered the floor and an acrid smell hung in the air. Another youth motioned to Brian where the two had burrowed through the debris. He was covered in dust and his T-shirt was streaked with blood.

"In there," he whispered, pointing to where they had burrowed. "There's two of them. One's gone, I think. The other is still alive."

Brian crept forward on his hands and knees into the hollow in the debris. The first figure was barely conscious, his face terribly lacerated and a look of terror in his eyes. Brian gave him the last rites quickly and then manoeuvred his way around the victim so that he would not have to move him. As he inched towards the second casualty, he could hear an ambulance siren in the distance. The second man lay facing into the debris. Brian gently eased the body onto its back. A wave of nausea swept through him as the body faced him. The whole left side of the man's upper body had been blown away. Half a face stared crazily up at him, blood seeping from the shattered skull. The left arm had disappeared and a smashed ribcage jutted through mangled, bleeding flesh. Brian fought desperately to control his heaving stomach. Somehow he managed to pray over the boy before turning away and vomiting over the pile of debris. He began to cry uncontrollably. He knelt between the two bodies uttering fragments of prayer. When the ambulance men reached him, such was his condition of shock they initially assumed he was a third victim of the explosion. They gently moved him out to the door. Before he did so, he stole another glance at the half-face of the dead man. Even in its horrific state, he was sure he could recognise it as belonging to Joe Rogan.

Rogan, May 19th 1971. Volunteer Joseph Pius, Kilcartin Upper. Aged 20 years. Killed on active service. Deeply regretted by his sorrowing wife Maureen, son Cathal, father, mother, brother, sisters, a large circle of friends and his comrades in North Monaghan Brigade PIRA.
"Ireland unfree shall never be at peace."

Brian tossed the newspaper aside in disgust. "Killed on active service," the death notice said. Blown apart trying to prime a bomb he knew nothing about. "Volunteer." How much "volunteering" had he done? Or had he been "asked"?

A huge cortège followed the removal of the remains from the County Hospital to Kilcartin. On the fringe of the village a lone piper stepped out in front of the hearse and led the cortège into the churchyard. Six men dressed in black and wearing balaclavas and berets flanked the hearse, three on either side, and accompanied the remains with a slow march. When they reached the churchyard, they stood to attention as Joe Rogan's immediate family removed the coffin from the hearse and bore it aloft on their shoulders. It was only then that Brian, waiting to greet the remains on the church steps, noticed that the coffin was draped with a tricolour with a belt, beret and black gloves arranged along the lid. The coffin was borne forward.

Brian raised his hand to halt the mourners. He took a deep breath and spoke aloud.

"This is a church funeral, not a military display or a political protest. I cannot allow these military trappings to be placed on the coffin while it rests in the house of God."

There was a tense silence before Charlie Rogan motioned to a bystander to take his place under the coffin. He then moved forward to stand directly before Brian. The skin went taut on his face.

"What are ye at, Father? I told you once before that the Rogans always paid their dues to church and chandler. This boy has paid his dues to the nation – the supreme sacrifice. No priest of Rome is going to deny him recognition of that!"

"All that's being denied is the display of military trappings in the church – "

"My son died a soldier's death and you're not going to tell me – "

"I cannot allow – "

Another voice cut across Brian – faint and hoarse but quite distinct in the tense atmosphere.

"Let them proceed, Father Johnston." Across the yard, in the doorway of the presbytery, stood the parish priest. A rug had been draped around his shoulders and he leaned awkwardly on a walking-stick.

Brian stared at him, incredulous. Father Reilly gave an impatient wave of his hand. Brian felt humiliated. He had no choice but to turn and ascend the steps, leading the coffin into the church with prayers.

At the funeral Mass next morning there was a palpable tension in the air as the congregation packed the church and spilled out in the churchyard. Brian had weighed carefully what he would say in his homily. He would have to make a stand.

"My dear people. Death is never a welcome visitor but when he comes to steal one so young, we feel we must cry out. We must rage and ask, Why? – Why? Why? Why? If one so young dies from a wasting disease or in a car crash, it is very difficult to find an answer. Impossible really. But when a young man dies in the manner in which Joe Rogan died, someone has to answer the question, Why? And for what?"

A ripple of nervous coughing echoed through the church.

"For what? I married this boy just two years ago to a quiet young girl. In this very church. She's sitting there now, not twenty years of age, mother of a young child with another on the way – and widowed. Is this fair? Is this right? I'm not putting those questions to God. I'm putting them to those who led Joe Rogan down the path that led to his horrific death. I'm asking what his death achieved – "

It began in the lower part of the church nave. Barely perceptible at first. A few feet stamping on the wooden floor. Slowly the noise grew as others, emboldened by the action of those around them, joined in the tattoo. Brian stopped and gazed at the impassive faces first in surprise, then in anger. He tried to continue.

"You may drown out my voice but you cannot drown the voice of your own conscience!" he shouted, then walked away from the lectern and stood bowed before the altar until the stamping eventually subsided. He steeled himself to begin the offertory of the Mass, but each time he gripped the chalice, his hands trembled uncontrollably.

Before he gave his final blessing to the coffin, Brian addressed the congregation once more.

"I wish to state that it was not my intention to upset Joe's family and friends. I too am upset at his needless death. Violence is not the way."

He could still sense antipathy and coldness in the graveyard and when the black-clad party stepped forward to fire a volley of shots over the coffin, he slipped away.

He avoided his parish priest in the days that followed. There was no point in arguing. He genuinely feared the effect it would have on the ailing man. On a number of occasions when Brian answered the telephone the only sound to be heard was a strange clicking noise. It was some days before he realised it was the sound of a safety catch on a revolver being released.

William Barden was interviewed on BBC television: "The IRA will not bully us into submission," he asserted. "We are the sons and daughters of Ulster – decent, hard-working and God-fearing people. We will not be intimidated into surrendering to a Rome-run state. We will defend ourselves as we see fit." Hannah nodded and made a little patter of applause.

"That was excellent, dear," she said quietly.

"Indeed I said a lot more than that but as usual they cut it. Too close to the bone for them – "

The window behind him shattered and Hannah watched, horrified, as the ball of flame spun upwards in an arc before crashing onto the fireplace. The resulting explosion sent tongues of flame snaking across the carpet. For a moment Hannah stood transfixed. Fleeting images shot across her mind – in a panic-stricken crowd, a faceless baby, a man carrying two greyhound pups. She began to scream hysterically.

"The police! Ring the police!" William Barden shouted to his wife as he began to flail at the flames with his jacket. Hannah seemed paralysed by fear.

"The police!" her husband roared again as he fought to control the flames. She finally sprang into action, raced out to the hall and phoned the police. She noticed the glow through the glass panelling in the front door. A second petrol bomb had hit the door. It had only scorched the paintwork but pools of flame still floated and danced on the doorstep. Hannah grabbed a vase of lilies from the hall table and emptied the vase on the flames.

By the time the police arrived, William had beaten out the flames in the drawing-room. Only the broken window and the scorched door bore testimony to the attack from the outside. The RUC inspector surveyed the burnt carpet in the drawing-room.

"I think in the light of this attack we should arrange some regular

336

surveillance of your house, Mr Barden," he said. "I will call around tomorrow."

Hannah stood in the kitchen shivering. She gathered her cardigan tightly about her.

"I think you should go down to Sloe Hill for a while until this business blows over," William said. "I'll join you in a few days. I have some business with Winston." Hannah nodded in agreement. William leafed through his bible. "And now we give thanks to the Lord for our deliverance from the hands of evil men and we renew our faith in the Lord. Psalm 91." He held the book aloft as they both knelt on the flagged kitchen floor.

"You shall not be afraid of any terror by night,
Or of the arrow that flies by day,
Of the pestilence that walks about in darkness,
Or the plague that destroys at noonday."

Hannah had a recurring image of her cousin's smiling face the night before the bomb fell.

"The Lord himself is your refuge:
You have made the Most High your stronghold.
Therefore no harm will befall you
Nor will any scourge come near your tent."

CHAPTER FORTY-FIVE

I T WAS THE FIRST FRIDAY IN AUGUST 1971. BRIAN JOHNSTON HEATED UP A bowl of soup and turned on the radio to hear the lunch-time news. He had spent a busy morning bringing communion to the sick. There was an edge to his appetite. He searched in the fridge for the makings of a sandwich. Half a tomato and some stale cheddar. Typical end-of-week fare. He sliced the cheese with some difficulty. The news headlines drifted past him – threat of internment in Northern Ireland – Dublin Horse Show – house under siege by RUC – a bomb in Derry. The tomato soup tasted good. Even the sandwich –

The newsreader's voice suddenly arrested his attention.

" – in Sloe Hill on the Northern side of the Monaghan-Armagh border. The RUC have now surrounded the house. It's understood that a number of gunmen entered the house at about nine o'clock this morning. Neighbours say they heard shots being exchanged inside the house. There are unconfirmed reports that the Reverend William Barden, leader of the United Church of Ulster movement, and his wife may be in the house . . . "

Brian felt a chill sweep through his body. The food suddenly became tasteless. He searched for a pen and some paper and wrote a hurried note for Mrs Martin, the housekeeper. He propped the note against the butter dish and left the house. Within five minutes of hearing the news, he was on his way to Sloe Hill.

He parked his car halfway up Caraher's Brae and negotiated his way on foot across the crude metal and concrete barrier. When he reached the top of the brae he was confronted by two soldiers, who leaped into his path from the roadside ditch.

"Halt! Out for a stroll are we, *padre?*"

"I need to get to Sloe Hill. I've come from – "

The smirched and streaked faces watched him intently.

"Oh we know where you've come from, but this is as far as you go."

"Look." Brian pleaded. "My sister is probably a prisoner in that house. I need to know what's happening?"

"Stand and face that tree! Legs apart. Hands on the tree," the soldier snapped. His colleague frisked Brian rapidly.

"Sorry about this, *padre,* but we don't even know if you are a *padre.* Now what's this about your sister?"

Brian repeated his request. The soldiers looked at each other.

"This way." They motioned him forward. When they rounded a bend in the road the scene changed completely. A huge army personnel carrier straddled the road. A row of Land Rovers lined one side of the road all the way up to the gates of Sloe Hill. Armed policemen lay against the hedge at intervals, guns trained on the house. The soldiers took him to the leading Land Rover where they spoke to the officer in charge. He was already shaking his head as he approached Brian.

"Look, Father. We have a siege situation here. This is no place for civilians – "

"I was born in that house. My sister is married to the Reverend Barden. My aunt is in there also. I need to find out if they're all right."

"We all need to find out. That's why we're here," the officer said tetchily. "The trouble is we don't know what we're dealing with. There are at least two gunmen. One's a nutter who's demanding safe passage across the border or – "

"Or what?"

"Shouldn't have said that." He paused for a moment. "Or they'll start killing off the occupants."

"You have to let me go in there, Inspector," Brian pleaded.

The officer shook his head. "Can't take that responsibility."

"What are you going to do? Wait for him to start killing them? At least I can buy some time, if nothing else. At best I can talk them out of it."

The inspector looked at Brian with growing curiosity.

"Wait here," he said before returning to the Land Rover. Brian could hear a muffled radio conversation from within the vehicle. He looked across the hedge. There was no sign of movement inside the house.

"OK. You can try negotiating with them but I have to warn you. You're putting your own life in danger. Those fellows are no boy scouts!"

Brian nodded. The officer moved to the gateway and shouted through a loudhailer.

"Attention! There's a priest coming through to talk to you. His name is Brian Johnston. He just wants to see if we can come to an arrangement. He's unarmed. There are no tricks. Is this all right with you?"

There was a long eerie silence, broken only by the yelping of a dog, locked away somewhere in an outhouse.

"Is this all right with you?" the officer repeated. The front door was opened slightly.

"All right. Let him come in with his hands on his head. Any smart tricks and he gets it too."

"No tricks. He just wants to talk."

The inspector nodded to Brian.

"You heard him. At least find out how many there are, where they are and how well armed – "

"You're not going to try anything foolish?"

"Of course not. Good luck."

Brian stepped forward, put his hands on his head and began walking towards the door. His heart pounded as he drew nearer to the house.

"Help me now, Mother. Help me now," he prayed silently. He pushed in the front door and stepped inside.

Immediately the door slammed behind him. He was grabbed roughly by the shoulder and spun around to face the door.

"Hands up against the door. Legs apart!" a voice barked.

"I'm not armed – "

"Shut up!"

Brian was frisked again.

"OK. Turn around." Brian turned to face his captor. A slight young man, very young to judge from his voice. His face was clad in a balaclava. The rifle was crooked awkwardly in one arm. The other arm hung limply by his side. A dark red stain covered the shoulder of his jumper.

"In there!" He motioned to the drawing-room.

The first face that Brian saw on entering the room was that of Aunt Rose. She sat in one of the fireside chairs, her hands gripping the arms of the chair tightly. She stared blankly at Brian and showed no sign of recognition. She was obviously in deep shock. From the other end of the room came a quiet sobbing. When his eyes adjusted to the dim light, he recognised his sister on her knees cradling her husband's bloodied head. He moved quickly towards her.

"What happened?" he whispered.

"They shot him," she sobbed. "He didn't even have a gun. Just shot him in the head. Never gave him a chance. He's dying."

William Barden's face was drained of all colour. Blood seeped through the matted hair above his right ear.

"Please ask them to let him go," Hannah cried. "Please."

"He can go as soon as we get our terms." The gruff voice startled Brian. In his anxiety to get to Hannah he had not noticed another balaclava-clad figure who had been kneeling over the prostrate figure of Winston Pollock. They had been hidden from Brian's view by a sofa. The gunman stood upright. He was very tall and lean. A three-quarter length leather jacket flapped loosely around his body. Brian noticed the handgun jutting from the pocket. In his hand the gunman brandished a long kitchen knife. Beneath him, Winston Pollock lay face down on the floor, his arms and legs firmly trussed, but seemingly uninjured.

The gunman approached Brian. "I hope those bastards haven't put you up to – " He paused and stepped closer, peering into Brian's face. "Jesus Christ, I don't believe it! It's Briny, isn't it? Jesus Christ! Briny!" He whipped off his balaclava. "Don't tell me you don't remember your old school pal, Briny?"

Now that his head was uncovered, Brian had no difficulty recognising the gunman. Even though he was practically bald, the angular features had not changed in over thirty years. Brian found himself looking into the face of Johnnie Caraher.

CHAPTER FORTY-SIX

"HELLO, JOHNNIE. LONG TIME NO SEE!" BRIAN SAID.
"Long time indeed! You haven't a fag on you, I suppose?"

"Afraid not."

"Christ, what a kip to break into! We came in looking for guns and these miserable gets haven't even a cigarette between them!"

"Johnnie, you're doing yourself no good here – "

"Shut up, Briny. The last fuckin' thing I want is a sermon."

"It's not a sermon, Johnnie. It's your old school pal – like you said."

"You were the only pal I had all right, Briny – but that was no school. It was a fuckin' prison. I hated every minute I was there – day and night. Do you remember Mother Caritas, Briny? If I had her here now I'd carve her into little pieces." He swung the knife around in an arc. "Fuckin' ould bitch. She made little of me every chance she got."

He ran his thumb lightly along the blade.

"Do you mind you showing me how to skin an Orangeman, Briny?"

"That was only a joke about peeling oranges – "

"No! No! I mind it well. Do you know how to skin an Orangeman? That's what you said to me, Briny. Well begod if we haven't the real thing here!"

He turned and approached Winston Pollock.

"This bastard Orangeman has caused us no end of trouble. Him and his cronies have killed and maimed several of our boys. It's about time he got a dose of his own medicine!" He held the point of the knife over Winston's cheekbone. "What was it you said, Briny? Get down under the skin – ?"

"No, Johnnie, no!" Brian pleaded.

His plea was ignored. With an eerie cackle Johnny sank the knife into Winston's cheek until it met the cheekbone. Winston roared in pain as blood spurted from the wound.

"Under the skin. Then peel away a strip – "

"Jesus, Johnnie, no!" Brian screamed as he lunged towards the crazed gunman.

The knife tore a horrible scar right down to Winston's chin. He roared both in pain and in rage at his humiliation. Johnnie turned quickly to ward off Brian's charge towards him. In one deft movement he overturned Brian and pinned him to the floor with his left arm while he held the knife poised above him with his right.

"Jesus Christ, Briny. You may be my school pal and you may be a priest but if you try anything like that again, I'll cut you fuckin' wide open! Wide fuckin' open! Do you understand?" His eyes stared wildly down at Brian, who could barely speak.

"All right, Johnnie," he croaked. "All right. I'm sorry."

The screaming and general commotion brought the second gunman running in from the hall.

"What the fuck is up, Johnnie?"

"It's all right, Damian. We were just having a bit of fun, weren't we, Briny?" Johnnie laughed as he released his grip on Brian. "You call out to those bastards outside, Damian. Tell them if we don't get news of a safe passage in fifteen minutes there'll be bits of Orangemen flung out that door at intervals – a few fingers, an ear, whatever's handy!" The young gunman disappeared again. "He's a good lad, Damian. This is his first mission, Briny. Christ! Has no one a cigarette?"

Hannah was now sobbing hysterically in terror at what she had witnessed. Brian crawled to her on his hands and knees and put his arm around her shoulder.

"Can you not shut that bitch up?" Johnnie snapped. "She's getting on my nerves."

"This is my sister, Johnnie," Brian said calmly.

"Your sister? You must be fuckin' joking!"

"No, I'm not. She's my twin sister."

"Well how did your sister end up married to that – ratbag?"

"These things happen, Johnnie."

"I was right about you, anyway, Briny. Didn't I say you'd be a priest? Didn't I?"

"You did, Johnnie – "

"Christ, I'm dyin' for a cigarette." He marched over to the door.

"Is there anything happening out there, Damian?"

There was no reply. He roared the question again.

"Nothing," came the weak reply.

"Are ye all right, Damian?" Johnnie shouted.

"I'm feeling a bit funny, Johnnie."

"Hang in there, son. We'll sort them out. You'll see." He turned back into the room. "Maybe a couple of Pollock's fingers will help them make up their minds."

He moved towards Winston who lay face down in a pool of blood.

"Please, Johnnie," Brian said. "There's been enough blood spilt."

"Enough blood? Enough blood is right. And most of it is ours. I'm tellin' you – I'll give those pigs outside five more minutes and by Jesus then we'll see some blood, won't we, Pollock – bollox?" He jabbed his boot into Winston's ribs.

Brian considered the scene before him. Rose rigid and immobile, a deathly pallor on her face; Winston trussed like a boar, emitting strange gurgling sounds as the blood from his slashed face seeped in an ever-widening stain into the carpet; William Barden barely breathing, slumped in the arms of his gently sobbing wife, Brian's sister. S-s-sister, s-s-sister, the train had said. The only mobile figure in this whole tableau of destruction was Johnnie Caraher who grew more agitated as the minutes ticked by on the ormolu clock on the mantelpiece. He muttered to himself and made repeated thrusting gestures with the knife.

"Johnnie?"

"Aye?"

"There are people in this room who have done you no harm, can do you no harm."

"It's not my fault, Briny. We came in here for guns. Our intelligence learned that this shitbag had a load of guns hidden away. All he had to do was hand them over. Instead he tries to be a hero and pulls a gun on us. Hits poor wee Damian. He panics and lets fly at yon." He nodded towards William, glanced at his watch and called out to Damian.

"Anything happening out there, Damian? Damian?" He peered out into the hall. Damian was slumped against the door. Johnnie ran up to him and shook him repeatedly. "Come on, Damian. Don't fall asleep on me now!"

Brian followed him into the hall.

"It's no use, Johnnie. He's unconscious. He's lost too much blood. He needs help – and quickly."

Johnnie paced up and down the hall, pausing each time to look at Damian.

"All right! All right! Go out there and tell them to bring an ambulance up here. With the doors open. No funny business or someone gets it. I mean that, Briny!"

Brian eased Damian's body away from the door and opened it cautiously. He waved his handkerchief and called out Johnnie's request. Almost immediately the reply came back on the loudhailer.

"OK. Coming through."

Brian and Johnnie carried Damian out to the porch. When they attempted to move William Barden, Hannah became hysterical again.

"Don't touch him!" she screamed at Johnnie. "Haven't you done enough to him?"

"It's all right, Hannah," Brian reassured his sister. "There's an ambulance outside. He'll be looked after." Hannah relented and, although eyeing Johnnie with suspicion, she held William's hand as he was carried to the porch. Rose would not move, would not even respond to Brian's pleading.

"Right," Johnnie said. "That's it. Tell them to clear off, Briny."

"What about Winston?"

"He stays. He's my only passport out of this."

"But he's – "

"He stays!" Johnnie snapped.

Brian acceded to his command and the ambulance sped away.

"You'll get me out of this cock-up, won't you, Briny?" There was increasing desperation in Johnnie's voice. "Many's the time you did before."

"I'll do all I can to help you," Brian said. He had a fleeting image

of an awkward gangling youth calling after him as he left Drumcran
for Christmas holidays – "I'll see you, Briny."

"Do you remember Big Malachy, Johnnie?"

The question caught Johnnie by surprise. He bit his thumb
savagely.

"Of course I mind him. He used to beat the shit out of me – "

"Do you remember what happened to him?"

"What do you mean?"

"He had to leave. His brother was killed – hit by a car. Do you
remember now?"

"It's coming back to me all right."

"And the morning he was leaving you ran up to him and gave him
your toffee bar."

"Me? I don't mind that at all."

"I do, Johnnie. I've always remembered it."

"Yeah, well. What's a toffee bar?" He moved to the window and
peered through the drapes. "Jesus! There's a fuckin' army out there."
He bit into his thumb again. "It was all right for you, Briny. You had
the brains. You always knew where you were going. You had it
easy."

"Not as easy as you think – "

"Youse were the fellows that mattered to Master McCartan. A right
savage bastard that fellow. 'Caraher,' he says to me one day – nearly
pulling my fuckin' ear off at the same time – 'Caraher, I see a great
future for you in the bank – the turf bank!' And he near laughed his
thick fuckin' head off." He shoved a chair angrily out of his way.

"Well I've got news for him. Two weeks ago I was in the bank
with two comrades. And we walked out of there with forty thousand
quid. What do ye think of that, Briny? What kind of a penance would
you give for that?"

Brian shrugged his shoulders.

Johnnie peered out again. Winston Pollock was emitting a series of
low groans.

"Christ, Briny. I'm rightly fucked, amn't I?" Johnnie said, sinking
with a sigh into an armchair.

"If you walk out of here now – "

"I'll walk straight into prison for twenty fuckin' years."

"Not necessarily – "

346

"But there's nothing new in that. Life's been a fuckin' prison." He stood up slowly. He suddenly looked very tired. "Will you come with me, Briny? For old time's sake?"

"Of course. Before we go we'll have to do something for this man." He moved towards Winston. "He's not looking too good."

Johnnie handed him the knife, with which Brian cut Winston free. He propped him up against an armchair. Winston was still conscious, though greatly weakened by loss of blood. Brian hurried to the kitchen where he found a towel. He wrapped the towel crudely around Winston's head in an effort to staunch the flow of blood. He paused briefly to rest his hand on Rose's arm. "It's all right, Aunt Rose. It's over now," he whispered. There was no reaction.

"OK, Johnnie. We'll go now."

"Throw your weapons aside," the voice on the loudhailer said. "Then put your hands on your head and walk slowly forward."

Brian signalled with some anxiety that everything was in order. They began walking towards the gate. Johnnie spoke with great rapidity.

"Has your aunt still got the pub, Briny?"

"No, she – "

"Christ, I'd murder a pint there this minute – and a packet of fags. I mind that summer we spent there. Your aunt was real nice to me. D'ye know something – that was the only holiday I had in my whole life. And all the games of football we had with – what was that fellow's name?"

"Hanratty?"

"Aye, that's him – "

A single shot rang out. Its deafening report startled Brian. Johnnie was pitched forward by the force of the shot. The back of his head was blown open. Brian stood momentarily frozen in shock before staggering forward to catch Johnnie in his crazy fall. As he knelt on the gravel cradling Johnnie's shattered head he was aware of police running towards him in a line from one direction. He looked back to the house. Winston Pollock lay in the doorway, collapsed on top of a rifle.

Johnnie's eyes barely flickered. His voice was an almost indistinguishable croak. Only Brian would have recognised the words he attempted.

"I'll see ye, Briny," he said.

347

CHAPTER FORTY-SEVEN

T HE "SIEGE OF SLOE HILL," AS THE TABLOIDS STYLED IT, WROUGHT
devastation on the lives of those involved. William Barden lapsed
into a coma, in which state he remained for almost two years.
Hannah visited him daily until the trauma of the siege and the stress
of her husband's condition eventually took their toll. Shortly before
her husband died she was diagnosed as having cancer. Moira took
her mother to her home in England where she spent her final years.

Winston Pollock was charged with the shooting of Johnnie
Caraher. And, while found guilty, he was given a suspended jail
sentence. Rose was sent to a nursing home, having suffered a stroke.
Winston, left on his own in Sloe Hill, became increasingly depressed
and drank heavily, thus alienating himself further from his
community. Damian was sentenced to ten years in prison.

Priest Friend Talks Provo into Siege Surrender, the tabloid headline
proclaimed. In developing the story and in photograph captions the
words "priest" and "provo" were drawn closer together. The day after
the siege a reporter and a photographer arrived in Kilcartin to do a
feature on "the priest and the provo". Brian slammed the door in their
faces. He had to get away for a while. He arranged for a fortnight's
cover and headed for the only place he knew which would give him
the solace he sought.

"Here he is – the hero of the hour," Larry Rooney exclaimed when
Brian walked in the door of The Willows.

"Don't you start! I've had enough of that!" Brian wagged a finger at
Larry.

"Oh! The reluctant hero too!" He embraced Brian.

"Shut up, Rooney, and pour the man a sherry – or maybe something stronger, Brian?" Susan ushered the visitor into the drawing-room.

"A sherry will do fine," Brian said.

"Don't mind me. I only work here," Larry said with a sigh as he poked his head into the sideboard in search of the sherry.

"What are you complaining about?" Susan said, winking at Brian. "The food is good, the drink is free and you get to sleep with the landlady!"

"Ah well, two out of three ain't bad, as Harry the Horse would say – ouch!" Susan gave a sharp kick to the rear that protruded from the sideboard. Larry backed out of the drinks press and stood up holding his bottom in mock agony.

"What's the matter, dear? Brain damage?"

Brian felt like applauding Susan's sharpness. Larry adopted a deferential tone.

"No, milady. I just wish to state that I have located the sherry but I have observed that all the sherry glasses are – ahem – unwashed. Perhaps milady forgot – "

"Indeed, milady forgot to remind you to wash them."

Defeated, Larry trudged towards the kitchen.

"Six to five, my eye!" he sighed. "The odds are lengthening all the time, Brian! There isn't a handy old curacy going a-begging over there, by any chance?"

Brian sank into the chair and stretched his legs out in total relaxation. He was among friends and in the midst of love.

Thirty minutes later a slight, freckle-faced girl of about twelve burst excitedly into the room.

"Mummy, you'll never guess – "

"Alison, where are your manners?"

"Sorry. Beg pardon for intruding – but Mummy, you'll never guess – "

"Alison, this is Brian, a friend of Mummy and Daddy."

The girl turned only slightly towards Brian.

"Hello. Pleased to meet you."

"Now, Alison, what will I never guess?" Susan asked solemnly.

"Pepper. He jumped the three-bar gate!"

349

"He did? Well, that's wonderful." Susan turned to Brian. "Pepper is Alison's pony," she explained.

"Pepper is the best pony in all the world. May I take a carrot for him?"

"If there's one left. Pepper is also the best-fed pony in all the world!"

Alison skipped towards the door.

"Alison! Manners!" her mother commanded.

The girl turned sharply. "May I be excused. Nice to meet you, Brian. Oh, you're a reverend!" She turned and was gone.

"You're truly blessed there," Brian said.

"Amen! The apple of our eyes," Larry replied. "Mother's manner and father's charm and intelligence . . . "

"Doesn't she rise well above the handicap?" Susan observed drily. "Another sherry, Brian?"

They sat down to dinner. "Tell me about Vincent," Brian said when nobody else mentioned the name.

"He's retired from active service," Larry answered. "Arthritis – in some severe form. They have a big name for it but it simply means he's wasting away. You'll get a bit of a shock. He's in a nursing home in Catherton. I'll run you over tomorrow, if you like. What's the home called, Susan – Notre Dame or something?"

"Notre Dame is what you blew Giles's school fees on at Kempton last week."

Larry winced at the memory.

"Painful?" Brian asked.

"Don't ask. Several fingers burnt! Serves me right. Don't ever take a hot tip from an Anglican clergyman," Larry warned.

"He was a lapsed Lutheran, which was enough warning to the innocent," Susan corrected him. "And the home is called Nirvana."

"Do you not think it's entirely apposite that I end up here – in Nirvana?" Vincent said with a faint smile. "The final goal, according to the Buddhists, in which there is neither suffering, desire nor sense of self! I'm not sure that I've eliminated the first of those," he added quickly.

Larry's warning had been timely. Brian had been taken aback by the deterioration in Vincent's physical condition. He had quite simply

350

shrunk and his slight frame was stooped at an angle that seemed to defy balance. The once-elegant hands that had coaxed delicate melodies from piano keys were now swollen and gnarled. Vincent's withered, contorted body reminded Brian of a dead appletree in Podgy Kernan's overgrown orchard. That tree had frightened him as a child, especially when he stood at his bedroom window on clear winter nights and saw its twisted, knobbled form silhouetted in the moonlight, assuming grotesque poses.

The nurse brought a tray of tea and biscuits.

"Just like old times," Brian said as he poured the tea. "All we need are Conor Maguire – "

"Did you know that Conor died?"

"No. I've never – "

"Heart attack at forty-five. I accidentally met one of his foremen some years ago. He left a wife and five children. At least they were well provided for."

"God rest him," Brian sighed. "He was a decent chap."

"That's the way of it," Vincent said as he poured the milk clumsily into his tea. "Mr Shakespeare, as usual, had it right. We are merely players upon the stage. We have our exits and our entrances. Right now I've shuffled off into the wings." He cradled the cup awkwardly between the misshapen hands. "But Larry tells me you have been very much centre stage – the great hero!"

"Not where I want to be," Brian said. "It can be a lonely place." He recalled for Vincent the events at Joe Rogan's funeral.

"We both seem to have been cast in the role of outsiders, Brian. For myself, I could never fit into the group, the team. Something which caused great pain to my father. He lived for football and to have fathered a son who had no great liking for the game and no great wish to participate was the ultimate shame. And he let me know that!" He looked away to the window. "He let me know that. If I had left the priesthood so that I could play football for the county it would have been a source of shame for everyone – except my father. He would have rejoiced at the return of the prodigal son . . . "

He raised the cup to his lips once again with difficulty. "The important thing is to be true to your role while you have that hour on stage. There's a big audience out there and they don't all think alike." He laughed suddenly, shaking his head slowly. "Oh, listen to me –

351

sermonising again. I keep forgetting I'm retired! Let's talk of other things – ships and shoes and sealing-wax!"

"And music," Brian added. "I brought you a present," he said, handing the bag to Vincent.

"Chopin Nocturnes!" Vincent read with delight from the cover of the LP. "What a joy! Thank you, Brian. Here, you must put it on the record player for me – to make our soirée complete!"

For the rest of that visit they listened in silence. They were in communion just as they had been as master and pupil over thirty years previously. That and subsequent visits were part of a definite healing process for Brian. To be in the company of this wise and gentle man, to share their enjoyment of music, their common memories, the long necessary silences – all of this was a powerful balm for Brian in the healing of hurt, both recent and distant. And the peace of those visits was leavened with the raucous banter of daily life in The Willows . . . Parting from such disparate friends was even more difficult than on the previous occasion but Brian also felt much more affirmed and resolute on his return to Kilcartin.

CHAPTER FORTY-EIGHT

MAY 1975. AS BRIAN RETURNED FROM THE EIGHT-O'CLOCK MASS, THE SUN had already burnt off the mist that had drenched Kilcartin valley in the early morning. Brian had woken at dawn after a fitful night's sleep and when sleep would not return he had dressed and gone for a walk. The mist clung to the hedgerows and cocooned Brian as he stepped briskly along the Fanagh road. Despite the mist there was heat in the morning, yet he felt a strange chill within him. Now, hours later, with the mist dispelled, the chill remained . . .

He barely milked the cup of tea and drank deeply. He fumbled with the toast in retrieving it from the grill and burned the back of his hand. What was wrong with him this morning? It did not augur well for an important day. He fished the egg from the bubbling water with a spoon and slid it shakily into the eggcup. He cradled the chipped and cracked eggcup in his left hand while still blowing gently on the now blistered right hand.

"I don't know what you want with that oul' thing," Mrs Martin had said repeatedly, "and a grand set of wooden ones hanging there on the wall."

As he walked back to the table a tingle shot through his left arm. The eggcup slipped from his grasp and fell to the tiled floor, shattering on impact. The egg rolled crazily around the floor through the slivers and shards of china.

Brian felt strangely weak and groped his way along the table until he could sink into the chair. He sat there, contemplating the fragments that littered the floor. There would be no repair job this time. He sipped some tea. The weakness gradually subsided and he rose to sweep the broken eggcup into a dustpan. As he spilled the pieces into the pedal-bin he remembered a young girl, with plaited fair hair, smiling as he proudly handed her an eggcup. His sister. S-s-ister, s-s-ister, the train had said.

As soon as he had distributed communion at the midday Mass, Brian slipped from the church and drove to Cullyboe. As the car turned from the church gates into the street the telephone rang in the presbytery. There was nobody to answer it.

He parked his car outside the cemetery. One-thirty. He had time to spare. He went through the side-gate into the cemetery and welcomed the shade of the tall yew trees. He prayed at the graves of Bridie, Pete and James before moving on to his mother's grave. He dropped to his knees on the grass verge and remained there for several minutes. "Be with me now, Mother. Especially today," he whispered.

He left his car at the cemetery and walked up the winding road away from the village. About a half-mile up the road a crudely painted sign marked "Peace Rally" indicated a right turn along a narrow grass laneway. Bunting was draped along the hedgerow on either side of the laneway. Two o'clock. He was still early. He strolled through the cool shade of the hazel thicket at Culvinney before turning on impulse across the fields towards the Fadden river.

The river's level had fallen considerably. It was now little more than a stream meandering through the rushy fields. Brian followed its path for some distance. He found himself conversing aloud with his mother. "Whereabouts did it happen, Mother? Somewhere along here? Aunt Bridie would never say. Would never come back here. What a tormented state you must have been in? No one to turn to. How you must have suffered . . . "

He stopped and took *The Garden of the Soul* from his pocket He flicked through the pages until he came to the fragile almost fossilised remains of a daisy. Without understanding why, he lifted the withered flower gently from the book and placed it on the surface of the water. It disintegrated slowly as it was borne away on the flow of the river.

He was quietly surprised by the size of the crowd that had gathered in the coomb beneath the Mass Rock at Coomataggart. Two hundred and fifty. Three hundred, maybe. They were still arriving. He would have been happy with half that number. A tall figure came forward to greet him. It was Tom Hughes, the parish priest of Cullyboe.

"You're welcome, Brian. Thank you for coming. We'll give them another ten or fifteen minutes. They're still coming!"

Brian leaned on the great flat-topped boulder and began to speak.

"Dear friends. I thank you all for coming to this sacred place on this lovely day. It's good to be back in Cullyboe, the parish in which I grew up. I want to compliment Father Hughes on coming up with the idea of a Peace and Reconciliation Rally and to thank him especially for asking me to speak here. We live in dark times and we wonder is there any hope for the future, any escape from the spiral of violence.

"In other dark days, our ancestors came to this place in search of peace and consolation. We do not come here today in any triumphal sense, with a backward look. For me, this place is full of ghosts. I was always afraid to come here as a child, mainly because of a story my Uncle James told me about a man called Jemmy Patchy. Jemmy was a veteran and a victim of the First World War. A shell-shocked lost soul, who wandered the roads around here long after the war was over, thinking he was still soldiering. A gentle, harmless character who did harm to no one. And yet fifty years ago, almost, he ended up on this Mass Rock with his throat cut. Sacrificed on the altar, you might say. Why? Because he innocently found some hidden guns and he became an easy target for those who sought blind vengeance for a local murder. The victim in that murder happened to be – " Brian paused and swallowed quickly " – Gordon Johnston, who married Sarah McKevitt – my mother. Both of them died tragically within a year of each other. You see what I mean about ghosts . . .

"But it's the ghost of Jemmy Patchy that really haunts me. An innocent, slaughtered probably by men who had fought beside him in the war. And here we are fifty years later still slaughtering each other senselessly.

"The past is all about us here. The past lives in us but we must not live in the past. It is time to lay the ghosts of the past. It is time to leave the past behind. It is in that spirit, forward-looking and full of hope, that we come here today to try to begin to forge a different future. Somehow. It will not be easy but we must try for the generations to come and indeed for the ghost generations of the past. Let us begin."